PLANTS, PEOPLE, AND THE PLANET

An Introduction to Horticulture

Preliminary Edition

NATHANIEL MITKOWSKI AND BRIDGET RUEMMELE

Bassim Hamadeh, CEO and Publisher
Michael Simpson, Vice President of Acquisitions
Jamie Giganti, Managing Editor
Jess Busch, Graphic Design Supervisor
John Remington, Acquisitions Editor
Brian Fahey, Licensing Associate
Kate McKellar, Interior Designer

Copyright © 2014 by Cognella, Inc. All rights reserved. No part of this publication may be reprinted, reproduced, transmitted, or utilized in any form or by any electronic, mechanical, or other means, now known or hereafter invented, including photocopying, microfilming, and recording, or in any information retrieval system without the written permission of Cognella, Inc.

First published in the United States of America in 2014 by Cognella, Inc.

Trademark Notice: Product or corporate names may be trademarks or registered trademarks, and are used only for identification and explanation without intent to infringe.

Cover image copyright © 2010 by Depositphotos / spaxiax; © 2011 by Depositphotos / Xunbin Pan; © 2010 by Depositphotos / Shengqi He; © 2010 by Depositphotos / Sergejus Byckovskis; © 2011 by Depositphotos / Xunbin Pan; © 2011 by Depositphotos / Dimitris Kolyris; © 2011 by Depositphotos / Kzenon; © 2011 by Depositphotos / Simon Bratt; © 2009 by Depositphotos / Evgeny Karandaev; © 2011 by Depositphotos / Denis Tabler.

Printed in the United States of America

ISBN: 978-1-60927-210-4 (pbk) / 978-1-62661-520-5 (br)

www.cognella.com   800-200-3908

contents

the purpose of this book — vii

1 the beginning is a good place to start — 1
- a what makes biology
- b atoms
- c molecules
- d bonds
- e carbon

2 the importance of plants and growers — 7
- a plants, plant products, and their uses
- b what is horticulture?
- c your great grandfather, the farmer
- d industrialization

3 growing plants the new old way — 13
- a it's not easy being green
- b sustainable plants, sustainable populations
- c the victory garden

4 changing the planet — 21
- a plants changing the planet
- b people changing the planet
- c work with and against each other

5 chlorophyll: the most important molecule on the planet — 27
- a the rise of chlorophyll
- b how plants make our food
- c where it all takes place

6 a plant genealogy — 37
- a evolution is how we got here
- b how to tell the plants apart
- c the way biologists talk
- d the plant families

7 the spread of the plants — 45
- a seeds are plants on the move
- b where seeds come from—flowers!
- c two plants in one
- d people moving plants

8 the parts that make up the plant — 59
- a meristems
- b it all starts in the roots
- c always growing upward
- d the antenna for the sun

9 making the plant work — 71
- a plants are chemical factories too
- b signals and plant hormones
- c moving water
- d responding to temperature and day length
- e secondary products

10 how to grow plants — 81
- a weather is what it's all about
- b not all soil is the same
- c to fertilize or not to fertilize, that is the question …
- d just one of many meanings for organic
- e wilting or washing away
- f hot or cold?

11 plants and their enemies — 95
- a plants get sick, too
- b the pathogens
- c pests and herbivores
- d fighting disease in agriculture and nature
- e injury

12 natural and artificial plant evolution — 105
- a natural plant adaptations
- b breeding plants for agriculture
- c centers of diversity

13 fruits: all in the rose family — 115
- a what is a fruit?
- b introduction to the rose family
- c apples and pears
- d peaches and almonds
- e plums and apricots
- f strawberries
- g brambles—blackberries and raspberries

14 more fruits: not all in the rose family — 125
- a blueberries
- b cranberries
- c grapes
- d nutty fruits (pecans, walnuts, pistachios, cashews)

15 we eat these fruits but we call them vegetables — 137
- a cucurbits
- b tomatoes
- c eggplant
- d peppers

16 vegetables mom said were good for us — 147
- a the starchy potato
- b salad staples (celery, lettuce, etc.)
- c the same genus and species: cabbage, broccoli, brussels sprouts, and cauliflower
- d peas and beans—good for us and our gardens

17 plants that add flavorful zing — 163
- a mustard and horseradish
- b mints—more than just flavors of gum
- c parsley, sage, rosemary, and thyme
- d basil—it's the pesto
- e bulbs we eat—onions, shallots, garlic, and leeks

18 the crops that cover most of our fields   175
- a corn
- b soybeans
- c wheat
- d rice
- e other small grains (oat, barley, rye) and alfalfa

19 things we don't eat ... usually   183
- a cotton
- b tobacco
- c flowers
- d nursery and ornamental plants
- e turfgrasses
- f forests

credits   191

the purpose of this book

plants play a critical role in our lives and our societies. Despite their importance, many people live their lives without giving much consideration to the plants around them or all the things that plants have given them. Our dependence on plants is complete, but we often don't given them a second thought. We live in a busy world. People's lives are more complicated than ever before. Our connections to each other are more extensive, our activities are diverse and our time more limited. With all the demands of living in a modern society, it is no surprise that people rarely stop to smell the roses. This book is about smelling the roses. This book is about stopping to think about plants and what they mean to us. Our hope is that this book will give the reader an appreciation for the way people use plants, how plants have affected our lives, and how we and plants have shaped the planet we live on. There are a number of things this book is not. Firstly, this book is a not an exhaustive treatise on plant science. This book is also not a manual on how to grow plants. And this book is also not a substitute for a good set of introductory biology classes. All of those things are important but this is not the place to find them. Hundreds of detailed books exist on these and many more plant-related topics. Not this book. This book is written for the beginner. The book is written for the non-scientist, for the lay person. Hopefully, the reader will come away from this book with not only more knowledge about plants but an interest in them, either professional or amateur. Either way, this book is a good place to start learning about people, plants, and the planet.

1

a. what makes biology

according to cosmologists (people who study the universe), the universe is not very old. In fact, it's only been around for about 14 billion years. While that may seem a long time compared to our lives, it's not so long if you thought the universe was infinite. Before the universe began, we assume there was nothing. But since no one was around, it is impossible to know what was actually there. When the universe did begin, it started out very small—a tiny little patch of infinitely dense material. That patch exploded outward, releasing all the matter and energy currently in the universe. Over time, all of that stuff cooled down and formed the galaxies, stars, and planets we see today. The universe is still moving, getting larger, as our galaxy moves further and further away from where the universe began.

So what does this have to do with biology, the term we use to describe the study of life? At the beginning, not a lot. When the universe came into existence, there was no such thing as biology because there was no life. The earth itself is only about 4 billion years old, and as far as we can tell, the conditions that create life (more on this later in Chapter 3), are only present on planets. But while there was no biology, there were chemistry and physics. These three disciplines—chemistry, physics and biology—form the cornerstone of any education in science. These are the three basic subjects students engage in (often grudgingly) that describe how our world and our universe work. But only chemistry and physics were there at the beginning. Biology only came about when life evolved. In order for life to evolve, you need three things: chemistry, physics, and carbon. In other words: Chemistry + Physics + Carbon = Biology.

This book is about life and biology. Consequently, we will not talk too much about chemistry or physics, but there is a small amount of these subjects that any reader needs to know, as do all biologists and horticulturalists.

b. atoms

The first thing any biologist must know before we talk about biology is the atom. Atoms are the smallest recognizable piece of any part of matter. Every atom is a type of element. Atoms are the smallest piece of matter that retains the characteristics of an element. A good example is gold. Gold is an element. There are many things that make gold special. Gold is yellow. Gold is a conductive metal (electricity will move through it). Gold does not rust or corrode. Gold is soft. These are some of the important characteristics of gold. And one atom of gold is still gold. It is infinitesimally small, but it is still gold. When you take billions and billions and *billions* of gold atoms and stick them together, you get something substantial, something you can see. For perspective, a penny weighs about 3 grams. Cut a penny into three equal pieces and you get 1 gram. One gram of gold contains about 3,060,000,000,000,000,000,000 atoms. Clearly, atoms are small.

Every element that has ever been discovered can be found on the periodic table. This table is a stylized diagram that groups elements into different categories, based on their intrinsic properties. The periodic table is based on work developed by Dmitri Mendeleev, a Russian chemist, in the 1860s. The table usually contains the abbreviation for the element (Au represents gold) and other information, including the atomic weight and atomic number (which is the number of protons) of each element. The table is called "periodic" because it is groups the elements into similar classes, called periods.

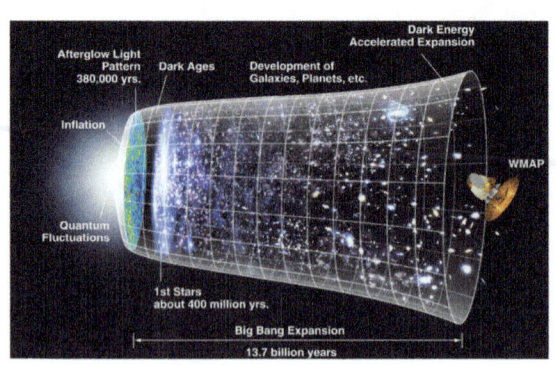

Figure 1.1. The Universe began with the Big Bang. As matter travelled across the empty void, it condensed and formed galaxies and planets. On these planets and on our planet, life eventually evolved.

If we were to look into an atom, we would get three main pieces: electrons, protons, and neutrons. The middle of the atom contains the protons and neutrons, stuck together like a loose ball that is call the nucleus. Protons are positively charged particles, which are critical in determining what element an atom will be. The number of protons in an atom gives an atom its qualities. Gold has 79 protons. If we could remove or add some of the protons, the atom would be completely different. If we added one proton so that we had 80, our gold atom gold would turn into mercury. Mercury is also a metal but at room temperature it is liquid, silver in appearance, and can be quite poisonous. But we can't practically add a proton to a gold atom. It is very hard to change the number of protons in an atom. The most successful way humans have done this is through a nuclear reaction (it's called a nuclear reaction because the nucleus is broken apart or joined together with another nucleus). In a typical nuclear fission that might happen when a nuclear bomb explodes or in a nuclear power plant and where things are broken apart, the process starts with uranium (92 protons) and ends with a large array of different atoms, including iodine (53 protons), barium (56 protons), cerium (58 protons), cesium (55 protons), and others. As just mentioned, protons are said to have a "positive" charge, much like a little magnet. And just like a magnet, different pieces in the atom with different charges are attracted to each other.

Neutrons are also contained in the nucleus but their role is less important. The number of neutrons does not always correspond to the number of protons (there are usually more neutrons in the nucleus than protons) and they do not determine the qualities of an element. However, different atoms of the same element can have different numbers of neutrons; these are called isotopes. Uranium often comes in two forms: uranium-235 (with 92 protons and 143 neutrons) and uranium-238 (with 92 protons and 146 neutrons). Because they are both uranium, they share the same characteristics, with one exception: uranium-235 can be used in a nuclear reaction, it can be broken apart; uranium-238 cannot be broken apart. In addition, it's much easier for atoms to gain and lose neutrons than protons. Neutrons get their name because they are neutral—they don't have any charge.

The last piece of the atom is the electron. For the purposes of biology, the electron is extremely important. It is smaller than a proton or neutron and has a negative charge. Electrons are critical to life and drive biological systems. While protons and neutrons sit in the middle of the atom at the nucleus and tend to stay pretty firmly in place, electrons whiz around the nucleus at the speed of light and can often be easily dislodged from the atom. When electrons are removed from an atom, energy is released, and those electrons and that energy can be used for all kinds of cellular processes. Electrons are a critical piece in what makes life happen.

b. molecules

As important as atoms are, they are pretty boring by themselves. They really don't do much. But when multiple atoms and combinations of atoms link together, they can create just about anything. Sugars are a great example. Sugars are important sources of energy. But sugar is not an atom, sugar is a molecule, which is a combination of atoms. There are many types of sugars, but one important sugar is glucose. When 6 carbon atoms, 6 oxygen atoms, and 12 hydrogen atoms combine together in a precise arrangement, you get glucose. And we write the formula that describes glucose as $C_6H_{12}O_6$ (where the C_6 is for the 6 carbons, the H_{12} is for the 12 hydrogen atoms and the O_6 is for the six oxygen atoms). We normally think about oxygen and hydrogen as gases. But when they combine to form glucose, the molecule that is formed is solid at room temperature. So when atoms do combine to form sugars, they create materials that are very unlike their constituent atoms. When small molecules combine to form larger molecules, they can also create new structures. When thousands or millions of molecules of glucose are linked together, they make cellulose. Cellulose is the primary component of plant cell walls and the majority component that makes up paper. So, in effect, a piece of paper is actually just a giant sheet of sugar, but because of the way the smaller molecules are assembled together, it neither looks nor tastes anything like sugar. However, it still contains all the energy that the original sugar molecules were built with, as is easy to see when we burn a piece of paper and the energy is released as heat.

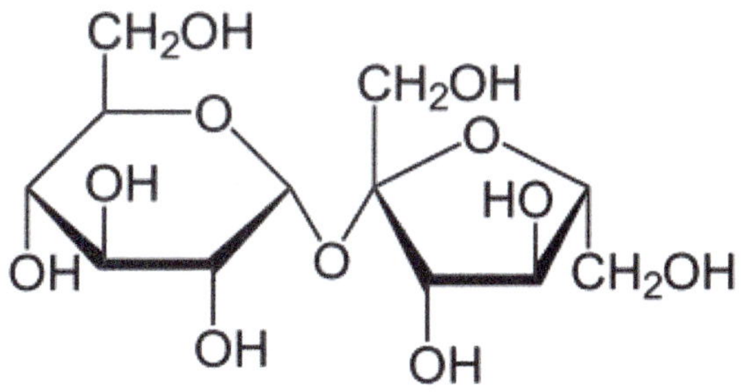

c. bonds

There are countless different types of molecules. However, the glue that attaches atoms together to form molecules is just as important. When atoms stick together (and atoms never stick together permanently), they are connected by bonds. A bond is actually nothing like glue, even though we often use this analogy. Glue is hard and rigid when it dries. What attaches molecules together is not an

Figure 1.2. Molecules are complex arrangements of atoms that possess characteristics very different from their constituent atoms. Sucrose, also know as table sugar, is produced by many different plants and is made from combining fructose and glucose.

actual substance like glue but an affinity for one another. An attraction. A married couple might spend their entire lives together, but there is nothing solid that connects them—their affinity for each other keeps them together. When atoms form molecules, they do so by either plain attraction (very similar to a magnetic attraction) or by sharing something important they both want: electrons. As was mentioned earlier, electrons are really important in biology. They are also really important in creating just about every piece of matter. There are three types of molecular bonds we typically talk about. The first type is a covalent bond. This is the type of bond where atoms share one or more electron. All atoms have electrons. But some atoms like electrons better than other atoms. When two atoms meet, the electron will occasionally jump from one atom to another. If the atoms share the electron, they become a molecule. That electron will whiz around the nucleus of both atoms, joining them together. This is the strongest type of bond.

If two atoms bump into each other and an electron is transferred but not shared, one atom will have too many electrons and one will have too few. Because electrons have a charge (like a little magnet), one of those two atoms will have more positive charge (fewer electrons) and one will have more negative charge (more electrons; remember, electrons are negatively charged). When a positive and a negative come in contact, they stick together. An ionic bond is formed when two differently charged atoms stick together. These bonds are weaker than covalent bonds. Covalent bonds do not like to come apart but ionic bonds will come apart very easily, often switching partners with other atoms. A good example of a molecule formed from an ionic bond is table salt. Table salt is made of two atoms, sodium (Na on the periodic table) and chlorine (Cl on the periodic table). When these two atoms meet, one of the sodium electrons goes over to chlorine. Now that the two atoms have swapped an electron, they have different charges and they stick together like little magnets. When you put table salt in

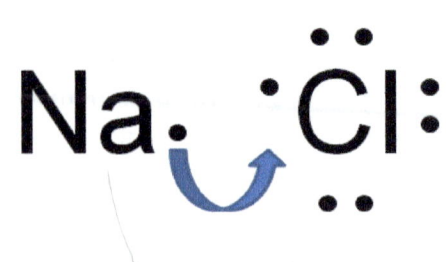

Figure 1.3. An ionic bond occurs when an electron is passed from atom to another, creating two atoms with opposite charges. The black dots in this diagram represent electrons available for sharing.

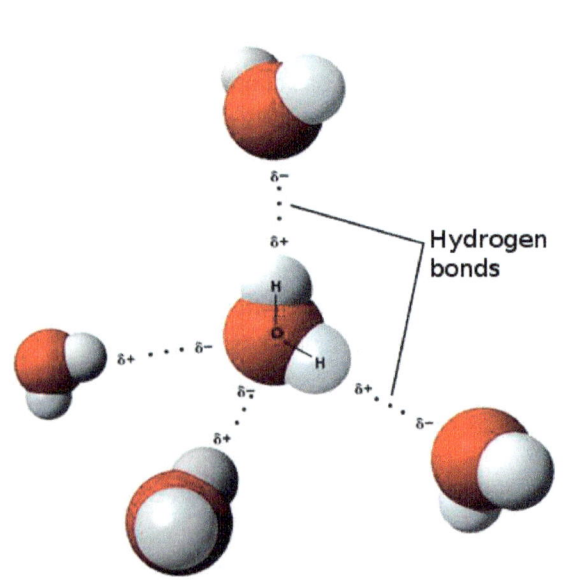

Figure 1.4. Hydrogen bonds are critical to life in Earth. The interactions between water molecules give water the special properties that living organisms, particularly plants, rely on to function.

water, it dissolves and the atoms (which are properly called ions because they have a charge) float around in the water, no longer always attached to each other. If you place a completely different salt into the water, those ions will intermingle with the sodium and chlorine ions.

The final type of bond we typically deal with is called a hydrogen bond. The hydrogen bond is a very weak bond, less like a bond and more like a loose association. In fact, a hydrogen bond does not occur between atoms in a molecule, it occurs between different molecules. If the covalent bond is like a marriage, the hydrogen bond is like a friendship. Sometimes, when atoms form a covalent bond, the electrons are not shared equally between all of the atoms in the molecule. A molecule of water is made up of one oxygen atom (O) and two hydrogen atoms (H) with a chemical formula of H_2O. While this is a covalently bonded molecule, the electrons get pulled to the oxygen side of the molecule more strongly than the hydrogen side. As a result, water acts like a mini-magnet. Instead of being a neutral molecule with no charge, the oxygen side of the molecule is slightly negatively charged from the presence of the electrons. On the hydrogen side, the molecule is slightly positively charged from the lack of electrons. The result is a molecule with different charges on it, called a polar molecule. Because it is a charged molecule, the whole molecule can stick to other molecules that are polar and have a charge. It's just like with little magnets. When two polar molecules stick together, the temporary bond they form is a hydrogen bond. It is called a hydrogen bond because hydrogen is usually the atom present that has a weak pull on its electrons and allows the molecule to become polar in the first place. These bonds are very important in determining how water will behave and how biological systems use water. Hydrogen bonds are critical to biology.

Figure 1.5. Carbon comes in many different forms. In this case, pure carbon is present in the form of graphite.

d. carbon

Carbon is an element. However, it is unlike any other element and has a number of properties that have made it an essential component to life. Carbon forms many different molecules and can occur as a pure element or in combination with other elements. A diamond is almost pure carbon. Similarly, the graphite in a pencil is also pure carbon. Carbon is also the primary component of oil and gas, forming the backbone of these molecules. And carbon is critical for life, forming the primary structure of most molecules in living things. Life on Earth requires carbon and is known as "carbon-based life."

So why is life based on carbon and not on other elements? Carbon's most important property is that one carbon atom can bond to four other atoms easily and simultaneously. This is important because it allows carbon to form a bridge, or backbone, to many different types of molecules and allows for many different formations and conformations. In addition to bonding to many other atoms at once, it can bond to a wide variety of different atoms like oxygen, sulfur, nitrogen, phosphorous, and even itself, carbon. And when carbon bonds to other molecules, it often forms gases such as carbon dioxide, which is required for photosynthesis. Finally, because of these properties, it can also form very long chains and rings of atoms. These large and complex molecules are responsible for life on Earth.

2

a. plants, plant products, and their uses

without the plants on planet Earth, there would be no people. Plants are fundamental to life on Earth, the survival of the human race (and all species), and the development of our civilizations. And whether we realize it or not, our lives revolve around plants and our future is in their hands. We are entirely dependent upon plants. Plants have shaped the world we live on and continue to shape that world. It is impossible to understate the ways in which we rely on plants every day and in every way.

A quick survey of the ways in which plants rule our lives can turn up both expected and unexpected results. First and foremost, we eat plants. All of the fruits, vegetables, starches, and grains we eat come from plants. But even the animals we eat are there because of the plants. Cows, pigs, and chickens all eat plants or products processed from plants. And while some animals are fed the byproducts of other animals, following the chain of consumption back to its origin would result in plants. Milk comes from cows but cows live on plants; milk is available for us because of the presence of plants.

Our processed foods all come from plants. Corn, soybeans, and other plants are grown in massive quantities and chemically or mechanically broken down to provide us with all the snacks and packaged goods we eat. High-fructose corn syrup comes from corn. High-fructose corn syrup is in much of what we eat, from soda to fruit punch to bread and fast foods. Other sugars come from beets or sugarcane plants. Coffee comes from coffee plants. Tea comes from *Camellia* plants. Aside from water, there is nothing we eat or drink that does not come from a plant or that does not have a plant somewhere in its genesis.

We even wear plants. Cotton is a major textile fabric and the clothing most people wear comes from cotton plants. Even if you choose to wear wool, that wool came from a sheep that ate grass. Without the grass, no sheep, no wool. We might think that polyester is an exception because some of it is a synthetic fabric. But some polyester actually comes straight from plants. Other polyesters are a type of plastic. And plastics come from petroleum, oil, and sometimes coal. But petroleum, oil, and coal all began as plant material, millions of years ago, that have been compressed and transformed into the products we see today. Because of this, aside from nuclear, water, or wind energy, our power plants that burn coal, oil, or natural gas rely on the products of plants. If your house is heated by one of these products, that heat relies on plants. If your house is heated with wood, it obviously relies on plants. If it is heated by electricity, how was that electricity produced? The majority (but not all) of our electricity is produced by burning fossils fuels, which are derived from plants.

Most of our electronic devices are made of large amounts of plastic. As we know, the origin of plastics will lead us back to plants. Anywhere plastic exists, plants were ultimately responsible, providing the crude materials that were produced hundreds of millions of years ago.

the importance of plants and growers

Plants are so important that we even feed plants to each other, to make more plants. One of the most significant advances in technology over the past 100 years has been the rapid increase in agricultural production. In order to feed the huge number of people on the planet, the amount of harvested product coming from each acre of land had to increase—there is only so much land available for agriculture. While there are a number of ways to do this and each method contributes a little to increased yield, the most successful approach is to heavily fertilize crops. The majority of the fertilizer used to feed the millions of acres of plants under agricultural production comes from oil and natural gas as a starter material. So plants that were alive eons ago are feeding plants today, which in turn, feed us.

It is not hard to see why humans and every other species of animal on Earth are so dependent upon plants. Plants play a role in all of our activities and endeavors. Without plants, there would be no people.

Figure 2.1. Horticulture means growing plants. From apples to wheat to sunflowers, anytime a person grows a plant they're doing horticulture!

b. what is horticulture?

The term horticulture can be summed up in two words: growing plants. More specifically, horticulture is the practice and/or science of growing plants. But as with most things, a complete definition of horticulture is much more complicated. Horticulture spans all the myriad species and varieties of plants. And while we often have some preconceived notion of what a horticulturalist is, many different people who do many different things can fit the description.

Some horticulturalists grow food crops. Horticulturalists make up a very diverse group of people who grow an enormous range of these types of plants. One grower may focus on only vegetables like tomatoes, watermelons, peas, potatoes, carrots, or lettuce. Or they may grow more exotic vegetables, like okra. In some countries, growers may produce crops like cassava, a staple food in many third world countries. Other horticulturalists grow fruit crops like raspberries or strawberries, crops that are very different from tree fruits like apples and oranges. Huge amounts of land in the United States are dedicated to field crops like corn, rice, wheat and soybeans.

As important as food crops are, many horticulturalists never grow an edible crop. Ornamental horticulture

We used to call horticulturalists "farmers." The term farmer, however, has slowly fallen out of favor and has gradually been replaced by the term "grower." This word seems to encompass more meanings than "farmer," which often conjures up the image of a guy wearing a plaid shirt, denim pants, red suspenders and riding a tractor. But whether you use the term farmer, grower, or horticulturalist, these people all do the same thing: grow plants.

is used to describe growers who produce flowers, shrubs, trees, and even turfgrasses. While these crops are often overlooked in the greater scheme of horticulture, these commodities contribute an enormous amount of value to local and national economies and employ hundreds of thousands of people across the globe. These crops may not be consumed, but they have a dramatic impact on the quality of people's lives and the quality of our communities. In addition to food and ornamental crops, some horticulturalists grow and manage crops with critically important uses like forest trees, used for lumber and paper pulp, cotton for clothing, and even corn that is principally grown for the production of ethanol, used as a fuel for our cars.

Defining horticulture based simply on the types of crops that are grown is only one approach that can be taken in understanding what horticulture is. Horticulture can be both practical and academic. That is, it can be an art, requiring years of experience and practice, or it can be a science where people examine the many different factors that favor or discourage plant growth and development. Horticulture is also a multidisciplinary term. Many different people grow plants and spend their time managing plants but may not consider themselves horticulturalists.

As important as plants are, the people who grow plants are just as critical to our continued existence. When people think about horticulture, plant science, forestry, or any other field related to plants, they often consider only the plant. But human societies rely upon farmers, horticulturalists, nursery owners, and many others to grow those plants and to bring plant products to market. There is a popular bumper sticker that says, "No farmers, no food." But in addition to food, those farmers grow crops that give us clothing and textiles, food required for the animals we consume, and many of the raw products that go into so many other parts of our lives.

c. your great grandfather, the farmer

Humans have many needs but some are very basic: food, water, shelter, clothing, safety, etc. Food is generally considered the most important. Without food, no one gets very far. While there are still many millions of people living in poverty, the majority of Americans and other people living in first-world nations have enough food to eat. If you are reading this book, you probably are one of these people. In fact, you probably have such easy access to food that it doesn't even occur to you that you could go hungry! The quality of this food varies, based on geography, economic status, and a bunch of other factors, but Americans have so much food that they throw away massive quantities of it every day. Despite this, many people around the world do go hungry. The guarantee

Figure 2.2. Even into the 1930's and 40's, small growers still managed their farms without the aid of machinery, as seen by a farmer in Beaverton Oregon in 1931. The full scale mechanization and industrialization of agriculture did not really begin until the 1940's and it was not until the 1960's that widely available fertilizers and pesticides made it increasingly efficient.

of food is a relatively new concept, associated with modern societies.

Today (especially in first-world nations like the United States) the production of food is undertaken by a very small percentage of the population. As populations have shifted from rural to suburban/city landscapes—and as the modernization and industrialization of agriculture has increased—fewer people have been required to produce food and fewer people have had an interest in food production. But this has not always been the case.

Your great grandfather may not have actually been a farmer, but many great grandfathers were. And even more great-great grandfathers were farmers. The production of food used to be the primary responsibility of every man, woman, and child in most families and communities. In North America, it has only been in the last century that this has changed. When America was colonized, there were no grocery stores. The only way to get food was to hunt it or grow it. While some food could be shipped from location to location by wagon or ship, transport across land or sea took a long time, if it happened at all. Perishable foods would often go bad in transport and they could not be stored well when they arrived at their destination. And even if a product could be transported, transportation was extremely expensive and the amount of food that could be transported was very small. Because early American colonists did rely on food from ship, their population potential was severely limited. Hunting generally could not support large populations for long periods of time either. Game animals would often be quickly hunted out or were simply not reliably present. And while there is no crop that provides a guarantee of success, farmed product can produce large amounts of food, many different crops can be potentially grown, and the resources to grow food products are generally present every year. Growing food can be sustainable over long periods of time (to some degree, given optimal conditions). As civilizations realized this, they became what we call "agrarian." That is, they farmed.

Because of the large amounts of food required by large societies, large societies set aside substantial portions of the landscape for growing food crops. If you drive through the northeastern United States today,

Figure 2.3. Industrialization allowed more crops to be produced and harvested more cheaply than ever before. A group of farm hands use an engine to power plant processing equipment in the 1900's in Rhode Island.

The Natural Resources Defense Council estimates that every year, Americans throw out almost 40% of the food they purchase at an estimated $2,275 per family, totaling $165 billion annually. Not only is this an enormous waste of food but at least 70% of the water Americans use goes into growing food, meaning were wasting enormous volumes of water. Where does all that wasted food go? Landfills. Less wasted food = smaller landfills.

you'll drive through mostly trees and forest, peppered with suburbia. But this was not the case 200 years ago. As more people immigrated and as populations exploded, land was needed for agricultural production. Not only was land required for crops but also for grazing animals. Today, most places that are completely forested were once open fields and farms. On a hike through any New England forest, you'll find miles of stone walls, defunct wells, and even the foundations of old houses and barns. Agriculture was everywhere. Every house was a farm and every community was full of farmers. Even those members of society that did other things for a living would still often have small plots and grow some food. Food and growing food went together: If you wanted to eat, it was up to you to make sure you were fed.

d. industrialization

As people spread from the East Coast into the central part of the continent, they built new farms and moved to places where soils were better, more land was available, and new opportunities existed. Agriculture and the need to produce food helped drive the expansion of populations westward. This general pattern continued up until the rise of cities. As the industrial age got underway, factories were established that built and produced many societal needs quickly and cheaply. No longer were items made by hand in the workshops of highly skilled craftsmen and metal workers. Tools, furniture, clothes, utensils, and hundreds of other products were mass-produced in factories. These factories needed to be located in convenient places so that they could ship and receive goods. Clusters of industry arose and cities developed, usually on rivers and coasts and other places where roads and travel were established.

The people required to run factories and populate cities had to come from somewhere. They usually came from either rural farms surrounding the cities or from immigrant populations that came to America without land or possessions, but with an ambition to succeed in a new place. As the cities boomed, the farms went into decline. Small farms were absorbed into larger farms. Other farms were abandoned. Today, there are large portions of the United States, particularly in the Upper Midwest full of abandoned farms, houses and communities. Children scattered and moved to the cities. As farming became mechanized and as pesticides and synthetic fertilizers became widely available, the number of

Figure 2.4. Early attempts at large scale agricultural pest control were crude. Here, a modest sprayer developed in the 1900's was used for spraying potato crops.

people required to farm thousands of acres at a time dwindled from hundreds to just a few. Land that was poorly adapted to farming was left abandoned or treated with fertilizers and improved agricultural/industrial practices squeezed additional yields out of every acre. Today, it takes only a handful of people to run a very large modern industrial farm.

The industrial age started globally in the mid 1700s with the Industrial Revolution and continued until about the 1970s in the United States. While factories and manufacturing still exist in the United States, they account for a much smaller portion of American activities than in the past. Because of high labor costs, environmental regulations, and the ease with which almost anything can be shipped, much of the industrial activity that used to occur in the United States has moved elsewhere. The United States is now a service economy. Our activities revolve around moving information, providing services, and developing new ways to look at and utilize these things. American efforts focus on health, education, financial services, and other knowledge-based activities. Consequently, fewer people today farm the land than ever before, potentially to our peril.

Despite that fact that food is critical to our existence, this critical resource is in the hands of very few people, located in very specific regions. The potential for a food catastrophe is great, something that would have been virtually impossible 200 years ago, specifically because of the lifestyle of the people at that time. In less time than the United States has existed, we have moved from a civilization of farmers to a civilization of service providers. The consequence is that few people consider how important plants are to our survival or have any idea of where our food comes from.

3

a. it's not easy being green

most industrial nations like the United States grow food in an industrial way. Huge amounts of land are managed by very few people using incredibly expensive and technologically advanced equipment. Farms that exist today are generally comprised of what used to be many smaller farms. As mentioned in the previous chapter, small farms were often abandoned or sold as they became less productive and less competitive. While even a large farm may be owned by a family, the farm is often subcontracted to a major corporate entity that dictates what products the farm owner will produce and how they must produce it. Industrial farming runs like any other industry, those crops that are most profitable or somehow critical to the industry are the crops which are produced. Industrial farming relies on large inputs of fertilizer, large inputs of pesticides, and often large amounts of water. Industrial farming is usually done in monocultures. In order to produce as much food as possible as cheaply as possible, the farmer often grows only one type of crop, and it is often a single variety or two—this is a monoculture. Industrial farming is the equivalent of using a factory to farm and most factories focus on one or a few products, just like an industrial farm. The biggest advantages of this technique are usually its low cost, its convenience for consumers, and the high uniformity of product. This type of farming took off in the 1960s as part of what was known as the Green Revolution, a period when agricultural output expanded dramatically through the use of newly developed technologies.

While every plant belongs to a species, such as "apple" (Malus domstica), there are many different varieties of apple that may have different colors, tastes, texture, etc. Granny Smith is one variety of apple, Fuji is another. Varieties are analogous to people from different countries- someone from Japan may be very different from someone from Norway but they are both people

Industrial farming works because of the society we live in and all the other technologies that have evolved over the past few decades. Before agriculture was industrialized, food products were generally produced locally and consumed locally, although there were some exceptions, especially in the colonial period when food was shipped across the Atlantic. Now, however, nothing we eat needs to be locally produced. Industrial farms can focus on any product they can sell, no matter where it needs to be sold. Because of refrigeration and interstate highways, the vast capacity for ocean transport, and even the regular use of airborne shipping, food products that are grown anywhere in a country or anywhere in the world can be shipped globally. We now have access to fresh tomatoes, strawberries, oranges, apples, and hundreds of other products—all year long. Each grower can specialize in particular products because there is a market somewhere for that product and it can make it to market in time. Plant products can be picked and shipped unripe and be ready for consumption on delivery. And plant breeding, genetic engineering and plant growth regulators have all greatly improved the quality of shipped produce.

Our civilization moves fast. As a result, we love fast food. And we eat enormous quantities of processed foods. The corn and soybeans and other plants that go into processed foods don't need to be kept fresh and ready for consumption. They can be dried, stored, frozen, or processed for incorporation into foodstuffs that would be barely recognizable by earlier cultures as actual food. The explosion of cheap processed foods in our diets has also pushed many farms into large-scale industrialization. Just like cars are put together from parts produced in factories across the globe, our processed foods are put together from ingredients grown processed, synthesized and chemically purified from all over the world and assembled in a factory.

However, there has been a significant amount of push-back in response to industrial farming over the past decade. In addition to the definite benefits to industrial agriculture, it has some serious negative effects. Most significant is the potential damage to the environment. Soils can be dramatically depleted and damaged when intensively managed; pesticides and fertilizers can leach or drift into the environment, damaging other ecosystems; groundwater can become polluted through pesticide use; and consuming large quantities of pesticides can be potentially dangerous, especially if a grower is not careful in pesticide application and monitoring. And it is even possible for industrial agriculture to speed the process of global warming. Some of these negative consequences occur no matter how large a farm is or how it is operated. However, the larger the farm, the larger the cumulative effects on the environment and society.

Why do Americans eat so much fast food (an estimated $110 billion, according to Fast Food Nation: The Dark Side of the All-American Meal by Eric Schlosser)? Obviously, the convenience of fast food is a major factor in their success but fast food is also extremely cheap and loaded with fats and sugars, two things that humans crave!

The Green Revolution has resulted in a substantial increase in available food. In some places in the world, it has prevented entire populations from starving, but there are other ways to farm and many of the opponents of industrial farming point to sustainable farming as a much more environmentally stable practice. What is sustainable farming? There are a lot of opinions and possible answers to this question, but sustainable farming is generally a collection of practices that can be utilized over long periods of time that require few, if any, outside inputs and maintain or improve the nature of the environment. That is, things that a farmer can do to farm indefinitely and independently while making his or her land a better place for farming. Sustainable farming incorporates aspects of recycling, conservation, and conscientious management to use the land without damaging it. Unfortunately, sustainable farming is not inexpensive. Sustainable practices can increase costs for both the grower and the consumer and sustainable farming relies on practices that work in the particular location in which the farm is situated. Not all sustainable farms are identical- there is no single recipe for a sustainable farm, these farms don't come out of a cookie cutter, and sustainable techniques require significant daily efforts from growers and substantial knowledge and skill.

It should be noted that while sustainable farming is usually organic, not all organic farming is sustainable. Organic farming is a term that means no synthetic chemicals, pesticides or fertilizers are used to grow a crop. Even without these types of materials, an organic crop can be grown through industrial farming.

b. sustainable plants, sustainable populations

The word "sustain" means to persist or last a long time. The goal of sustainable farming, therefore, is to grow food in one location for a long time. Most farming is not sustainable. Even our great grandfather farmers often did not practice sustainability. They would grow crops in one location for a long time and then their yields would drop off. They could no longer grow food in the same place, so they would move to a new location. The soil was said to be "played out" and could not produce a good crop. What was really happening was the depletion of important soil nutrients—organic matter was lost, erosion occurred, and by the 1930s, large areas of land from Iowa to Texas were in the midst of a dust bowl, brought about by poor farming practices and drought. One of the reasons why so much of the Amazon rain forest is continually cut down is so people can use the land for farming and animal grazing. When they play out the soil, they move to another part of the forest and cut that down for their next farm. Most played-out soils could have continued to be productive, however, had outside inputs like fertilizers, pesticides, and water been applied. In any event, they still would have been poor soils. The key practice in maintaining a sustainable soil is to minimize or eliminate the use of outside inputs (that is, those things that don't already exist or are produced on the farm). The fewer outside inputs you use, the more sustainable your farming practice.

As a result the Dust Bowl, millions of Americans left the plains states and hundreds of millions of acres were denuded of topsoil or severely degraded, all as a result of poor farming practices combined with extended drought.

Sustainable farming is important because industrial farming will fail. It is not inevitable that it will fail, but it will fail if it continues to rely on fossil fuels. Fossil fuels are any product that was produced through the decay and transformation of organic matter (plants) over the past few hundred million years. This includes oil, natural gas, and coal. Because fossil fuels take millions of years to produce and because there is a limited supply of them under the ground, we will eventually run out of them. In fact, it is estimated that half of the oil under the Earth's crust was used up by 2005. And unfortunately, the first half was easy and cheap to get at. The second half will be hard and expensive to pull out of the ground. Either way, though, we will run out of fossil fuels in the next 100 years or so. When this happens, most of the farming on the planet will cease if it still relies on gasoline and cheap synthetic

Figure 3.1. Industrial farming maximizes the amount of crop or animal yield in the smallest possible space. Crop plants are usually jammed together in the tightest rows possible. In the process of factory farming, animals are also packed together as tightly as possible.

fertilizer. If industrial agriculture fails, there will be a food catastrophe, and hundreds millions of people will likely starve. Although the depletion of fossil fuels is eventually inevitable and this is a bleak picture of the future, the future is not inevitable. The things we do now to positively impact our populations, our agriculture and the planet can stave off catastrophe.

Sustainable farming is intended to rely very little upon fossil fuels. Sustainable farms are generally small. They hark back to the original farms of the American 1700 and 1800s, where a few hundred acres (or less) were owned and managed by a family. Sustainable farms incorporate many different crop types in one location, rotated and managed for their natural strengths. Soils on sustainable farms are carefully stewarded, and the land is carefully managed. Animals are incorporated into sustainable farms. Animal waste is used carefully on sustainable farms. And sustainable farms market locally, they do not ship their products hundreds of miles by truck, train, or boat, which would require fossil fuels. Sustainable farming will not solve every problem. A farm in Ohio cannot produce strawberries for its customers in February, but it can produce melons in the late summer and fall with the use of other technologies like high tunnels and adapted varieties.

Figure 3.2. Poor soil management and other poor agronomic techniques combined with record drought led to the Dust Bowl in the Great Plains during the 1930s. Many farmers abandoned or lost their farms following repeated wheat crop failures.

Sustainable farming also falls short because it cannot produce the same amount of food as industrial farming. This is bad news for a planet currently undergoing a population explosion. The planet Earth has a certain "carrying capacity"—in other words, only so many people can live on the planet and expect to get fed. This capacity is not always the same. It is dependent upon the resources currently available. With fossil fuels, the resources today are greater than they were in 1800, even though there are about six times as many people now than there were then. In fact, without fossil fuels, many predict that the Earth's population would have begun to crash in the 20th century. Despite this, it is believed that there are currently 50 percent more people on the planet than it can adequately support. But when all the fossil fuels are gone, the carrying capacity of the planet will plummet, and this does not even account for global warming and sea-rise effects. This will result in a food catastrophe. There is always a possibility that technology will advance enough to help offset the lack of fossil fuels, but it seems unlikely that any alternative supply of energy that can be used to produce food will be as efficient as fossil fuels. And even most sustainable farming relies on some energy, often fossil fuels. Sustainable farming is a start but far from a panacea.

In order for sustainable farming to work, it needs to be integrated with sustainable populations. Sustainable farms are limited in their potential to produce food, and the populations that surround them will have

Figure 3.3. Throughout the Dust Bowl period, giant dust storms would move across the plains blotting out the sun and making breathing difficult. These storms reshaped the landscape and removed the valuable topsoil from affected farms.

to respond to less food. Cultures may change, especially in developed nations where few people farm. Future societies will likely have many more farmers and many fewer service providers. Processed foods will probably be less commonplace, a fact that will almost certainly result in less obesity and healthier diets. And it will be of the utmost importance that civilizations protect and carefully manage their farms and forests, for their use and the use of future generations. For many years before the Green Revolution, scientists and philosophers predicted famine, disease and the crash of human civilizations. These dire predictions were all based on the agricultural potential of the times in which these philosophers lived. Using their assumptions, their predictions were inevitable. We can only make predictions based on assumptions that we believe to be true when we make them. Based on current knowledge and current technologies, both energy and agricultural production on Earth are limited and we are drawing close to that limit.

c. the victory garden

Since the start of the 20th century, the United States has been relatively secure in its ability to produce food. The most recent example of a severe, long-term food shortage was during World War II. During that war, the United States put all of its efforts into winning the largest conflict ever fought in history, while doing it on two major fronts. Aside from Pearl Harbor, this war was not fought on American soil, but every American fought the war. Boy Scouts and community groups collected scrap metal for tanks, equipment, and ammunition. Women worked in factories to build war machines (unheard of in the 1940s). The production of cars for the general public was suspended so tanks and planes could be produced. Food and fuel were rationed to the public so that they could be diverted to the war effort.

While the idea originated in the First World War, Americans were urged to grow their own produce whenever possible during World War II. Gardens sprang up across the United States, in backyards, front yards, rooftops, and parks. They were called victory gardens because they would help the country achieve victory in the war by reducing consumer consumption of commercially grown produce. This would ultimately lower the prices on that produce and make more available for the War Department. These gardens did, in fact, have a significant impact on food production in the United States, yielding almost 50 percent of the produce consumed by citizens and increasing total agricultural output.

Figure 3.4. As suburban populations continue to grow, housing developments often encroach into agricultural areas, changing the character of farming communities and turning former crop lands into lawns and neighborhoods.

Recently, there has been a renaissance in locally grown food and the victory garden. As people begin to think about where their food comes from and how it is grown, garden plots, community gardens, CSAs (community supported agriculture), and co-ops are popping up in many places. As it stands now, we have plenty of food at hand, and these efforts are fun and rewarding but not really critical. However, when more people know how to grow food, the quality of food increases and the demand on food supply chains diminishes. No American needs a vegetable garden (yet), but the quality of freshly grown and fresh consumer produce cannot be surpassed. There are many reasons for why commercially grown food is of lower quality than locally grown food, but it usually boils down to unripe picking, refrigeration during transport, and the types of plant varieties used to supply national and international markets. All of these things reduce the quality of the food we eat. In addition, the age of produce is difficult to determine when it has been shipped and has sat in a grocery store for days or weeks.

Besides better taste, texture and overall quality of locally grown food (and it can't get much fresher than when it was picked an hour ago from your garden or from the farm down the street), locally grown food generally has had fewer pesticides applied. Industrial agriculture works because it is cheap. Food products are treated with chemicals to kill pests and stop diseases. Even when people are involved in the process of industrial farming, their role is often limited and their salaries are low. People are expensive. Mechanization is ultimately less expensive. When a grower sprays 1000 acres with a pesticide, there is an economic return that makes the pesticide economically feasible. Local growers, on the other hand, avoid pesticide applications because their clientele generally disapproves of these applications, and on an acre of lettuce, a pesticide application (or five) may be more expensive than the total worth of the crop. Because small growers usually plant a diverse selection of crops, the loss of a single crop among 20 different crops is unfortunate but not catastrophic. Luckily, most diseases and insect pests are limited in the plants they can damage (as are diseases and pathogens in the animal kingdom) so a grower may lose all of their tomatoes to a disease, but everything else will be fine. Finally, most homeowners who grow a vegetable garden rarely need to use pesticides. But when they do the amount they apply is small and has a minimal impact on the environment, compared to the significant amount of pesticide delivered to commercial crops on a commercial scale.

In the past 100 years, agriculture technology has developed rapidly. Unfortunately, many of these developments have been based on the availability and dependability of fossil fuels. As cheap energy becomes less available, people will have to adapt to a different type of lifestyle and food production. Food security will reemerge as a major issue for all people.

Figure 3.5. During World War II, Americans grew victory gardens as a way to offset food shortages in the country because of all the food being sent overseas for the wartime effort. Today, people grow similar gardens for many reasons including better quality, lower cost and more variety.

4

a. plants changing the planet

as we'll learn in detail in future chapters, plants changed the planet over the course of billions of years by flooding the Earth with oxygen. This oxygen allowed aerobic life (live that requires oxygen) to develop and thrive. But the plants changed the Earth in other ways, too. Related to oxygen, the plants allowed for the formation of ozone. The oxygen we breathe contains two oxygen molecules (O_2). This form of oxygen is very stable and very prevalent; about 21 percent of the Earth's atmosphere is oxygen. As an aside, this means that every time you breathe, you are inhaling a lot of components in the air that your body cannot use. Another form of oxygen does exist on Earth, however, called ozone. Ozone has three oxygen molecules (O_3) and is formed when intense solar radiation breaks apart an O_2 molecule (into one oxygen atom), which then combines with existing O_2 molecules to form an O_3 molecule (1 + 2 = 3). Unfortunately for us, ozone is extremely toxic. Who would have thought? One extra oxygen atom is lethal! Not only will too much of it kill you, but it can also damage plants. Ozone is often seen in small amounts in the summer when temperature and humidity get high. Ozone is formed in these conditions when O_2 combines with pollutants from cars, power plants, and factories. When ozone is formed at ground level, it can cause breathing problems (especially for individuals prone to asthma) and contribute to additional air pollution.

Despite the fact that ozone can be poisonous, we would not be here without it. The greatest concentrations of ozone typically occur high in the atmosphere, in what is called the stratosphere—30,000+ feet high. The ozone in the stratosphere actually protects all the life on Earth by absorbing harmful ultraviolet radiation from the sun that would damage DNA and slowly extinguish all living things. The sun produces a vast amount of energy. We know that plants absorb this energy as specific wavelengths of visible light and convert it into food. However, the sun also produces other wavelengths of light containing dangerous types of energy. One of these wavelengths is ultraviolet light. A little UV light is not harmful. In fact, humans require some exposure to UV light. When UV light reaches the skin, it stimulates your body to produce vitamin D that is then used throughout the body. But excessive UV light is very damaging. When you receive a bad sunburn, the burn is caused by too much UV light. If there were no ozone in the stratosphere, so much UV light would penetrate to the surface that everything would be getting a lethal sunburn all the time. Nothing could survive. For this reason, the holes in the ozone layer around the north and south poles are cause for great concern. Those holes were formed when CFCs (chlorofluorocarbons) were released into the air and eventually, years later, migrated to the poles. When they reach the ozone layer, they consume ozone. For this reason, CFCs have been banned in much of the world with the hope that the ozone layer will recover in the next few centuries and the ozone holes will shrink. Plants created ozone as they evolved oxygen and that oxygen migrated into the upper atmosphere. People destroyed oxygen as we created and evolved CFC's .

In addition to oxygen, plants have shaped the planet by their very presence. When plants establish themselves on land, they hold the land in place. Without root systems and canopies to slow the movement of water and hold down the soil, huge amounts of land would wash into the oceans. This is very easily seen in many places where large-scale agriculture is under way. When a crop is produced, the crop holds the soil in place. However, if the crop is harvested and no replacement crop is established (called a cover-crop) for the nongrowing season, fall and spring rains will wash large amounts of soil into rivers and other watersheds and eventually into the ocean. Some plants can even build new landmasses, as is the case with mangroves. Mangrove trees establish themselves in swamps and on the edges of estuaries. As they grow and spread, they actually increase landmasses and moderate the action of storm surges, protecting coastal and inland environments. Over the course of geologic time, volcanic activity and tectonic movement will push up new land masses, but the surfaces would look very different without the influence of plants.

Figure 4.1. While ozone is critical to filter UV light and protecting life on Earth, too much ozone in the lower atmosphere can be dangerous. Even plants, like this tobacco plant, can be seriously damaged by high ozone levels.

b. people changing the planet

The planet goes through cycles of temperature. These cycles are affected by many things. One major influence on planetary temperature cycling is volcanic activity. When large volcanoes erupt, they spew massive amounts of ash and chemicals into the air that can block out sunlight and cool the planet. These cooling phases can last for many years. The activity of sunspots (essentially storms on the sun) can also influence the climate on the planet by increasing or decreasing the amount of heat/light that reaches us. And the Earth itself is slightly tilted: The tilt of the Earth fluctuates (or wobbles back and forth) over thousands of years. These all combine to result in ice ages and ice-free, or interglacial, periods. Further, it turns out that once a cold or warm period on the planet occurs, it can magnify itself. If you have a lot of ice on Earth, the ice will reflect sunlight and keep the planet cool, which intensifies the cold and keeps the planet cooler longer. This is called a positive feedback mechanism, meaning that when a thing happens, it makes more of that same thing happen. This process occurs without human help and is considered a "natural process."

But people can also influence the cycles on the planet. We mentioned the Green Revolution earlier and how it impacted farming practices (by allowing for huge amounts of food to be produced cheaply and easily), but the Industrial Revolution has also dramatically impacted the climate of the planet. In order for

Figure 4.2. People change the planet in many ways, both big and small. When new roads and highways are established, landscape patterns are often altered. This can result in new ponds and wetlands where forest existed previously.

people to become industrialized, they need energy. The energy is needed to power machines that make possible industrial activities. The earliest source of energy for machines was wood or coal. As wood and coal were burned, they released heat that could be used to melt and refine metals or operate steam-powered equipment. Unfortunately, burning fuel to drive industrial activities can impact planetary climate.

Coal tends to be a better source of energy than wood because it holds a large amount of energy compared to its size. Coal is also very easy to get: You dig a hole or a mine in the right place and pull the coal out of the ground. It can be as easy as strip mining, where entire mountain tops are removed to get the coal or it can involve deep underground mineshafts. There are many variations of coal, but all coal is a type of rock that started out as some kind of living thing millions of years ago. Many people believe that coal and the other fossil fuels came from dinosaurs. Because the word "fossil" is used to describe these fuels, it is not an unusual

Figure 4.3. Although CO_2 and methane are invisible in the environment, when pollutants produced from fossil fuels combine in the form of smog they can easily be seen in the atmosphere.

conclusion to make. However, most of our fossil fuels are derived from plant materials. While animals of every shape have walked the Earth for hundreds of millions of years, far more of the organic material that has ever existed on Earth originated as some form of plant material. As these plants, algae, and phytoplankton died, they were covered by soil and rock over many years. Over millions of years, as the organic matter stewed under intense pressure of all the weight on top of it, it was converted into coal, oil, and natural gas.

Fossil fuels contain more energy than just about any other source of energy, making them excellent for any process that requires combustion. As industrialized civilizations moved away from steam power and toward internal combustion engines (the engines in cars, trucks, and boats) oil and gasoline became the premier fuel sources. In the early development of internal

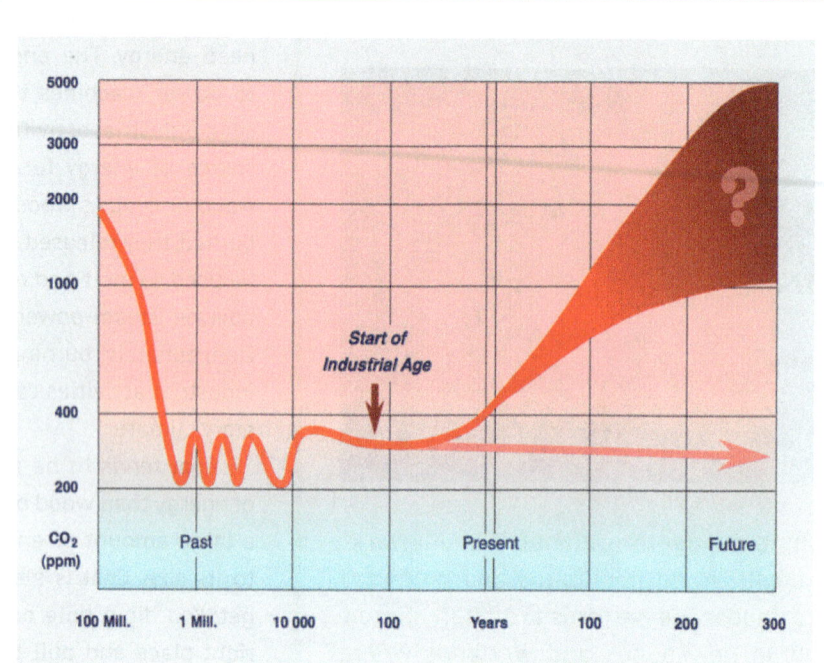

Figure 4.4. Although greenhouse gases were very high many millions of years ago (along with global temperatures), they have been relatively stable for the past million years. Since the Industrial Revolution and the burning of fossil fuels, CO_2 levels have started to climb dramatically.

combustion engines, ethanol was considered the most likely fuel. But ethanol contains less energy than fossil fuels, and production of ethanol met with a number of other issues that made it less desirable. But there are two significant problems with fossil fuels. First, it takes millions of years for them to be created and therefore only a limited supply exists. Once all the fossil fuels are used up, we cannot replace them. Second, burning fossil fuels puts a lot of extra carbon dioxide (CO_2) into the atmosphere. And when carbon dioxide builds up in the atmosphere, it raises the temperature of the planet. As we know, the temperature of the planet changes of its own accord but human activities can affect natural changes significantly. In most cases, the things people do on Earth actually increase global temperatures and as the planets temperature increases, many different things happen to our climate and our geography, few of which are good for us or the planet.

Many growers produce corn for ethanol production. Just as beer and wine are produced from fermentation, so is the ethanol used for fuel. Unfortunately, because ethanol has less energy per unit than gasoline, the amount of miles a car can drive on one gallon of ethanol is less than how far a car can drive on gasoline. Therefore, even though the cost for ethanol fuel (called E85 because it contains 85% ethanol and 15% gasoline) is less than pure petroleum gasoline, it costs about the same amount of money per mile driven. In fact, per mile, E85 may even cost you more!

Not only does carbon dioxide increase global temperature but so does methane. Methane is produced by not only burning but from natural decomposition in places like landfills, animals such as cows and sheep and the release from frozen soils in northern climates. As global temperatures rise and frozen soils melt, methane will become an even bigger problem as another positive feedback loop engages. Nitrous oxides (NO) are another gas that can increase global temperature. These gases are produced from burning fossil fuels but a significant portion comes from fertilizers. The most significant source of NO gases in the atmosphere comes from growing crops (see- another plant/planet/people connection!). When fertilizers are applied to the soil, some portion of the nitrogen, the most important part of the fertilizer, will turn into a gas and escape into the air. The EPA (the Environmental Protection Agency) estimates that 70% of the NO released into the atmosphere comes from agriculture. The very process of farming using industrial methods raises global temperature. And these are just some of the most common and well known gases that can increase global temperatures. Many other gasses exist that singly contribute only a small amount to increased temperatures but in combination can have significant effects.

The process of increasing the temperature of the planet through human activities is called global warming. Global warming can be directly linked to increased human industrialization, industrial agriculture and the consumption of fossil fuels.

c. work with and against each other

The planet we have today is not the same planet that existed hundreds of millions of years ago. One of the major changes in the history of the planet has been the temperature of the Earth and the amount of carbon dioxide and other gases in the atmosphere. These two are directly related. The more carbon dioxide, nitrous oxide and methane in the air, the warmer the temperature of the planet. These gases act like a one-way filter. They allow heat into the atmosphere, but not *out* of the atmosphere. For this reason, carbon dioxide is said to create a "greenhouse effect." When the Earth was much younger, there was a lot of carbon dioxide in the air, which meant the planet was much warmer. As plants consumed that carbon dioxide (or "fixed" it), they locked it up in their tissues and in the soil. When the plants were buried, the carbon dioxide was locked up even further and was eventually converted into fossil fuels. The Earth got cooler under natural processes and stabilized itself at those cooler temperatures. When we burn fossil fuels, we reverse this process that has been going on for millions of years. We, on the other hand, are warming the planet and destabilizing it. This also is a positive feedback mechanism. As the planet gets warmer, more permanent ice melts. When the ice melts, it can no longer reflect heat back into space. The exposed soil absorbs heat and makes the planet even warmer. As permafrost melts (permafrost is comprised of very cold soils in tundra landscapes that are almost always frozen and contain slow-growing plants), plant activity speeds up, and all the carbon dioxide and methane that was stuck in the frozen ground is released. And it is not just the gases produced from driving cars that increases global temperature. We do it in hundreds of ways each day like, burning coal or oil for electricity, refining oil, producing plastics, raising cattle, fertilizing crops, throwing garbage in the dump, and many, many other ways. As we live our lives we increase global temperature and the planet responds to our increases but putting forth its own increases.

The activities of people and plants are combining to rapidly increase the level of carbon dioxide in the atmosphere, which will ultimately raise the temperature on the planet dramatically. As glaciers and polar ice melt, sea levels will rise. Most of the human populations live in low-lying areas, close to the ocean. The population centers will have to move or be submerged. Hundreds of millions of people may be displaced. In addition, it is predicted that global weather will become more extreme, with more snow in the winter and more heat in the summer. Areas that once had significant precipitation may become arid and drought prone. In many areas of the United States, extreme weather patterns are becoming more commonplace. Severe hurricanes, tornados, snow storms, hail storms, drought and fire are a regular occurrence. While these events happen every year, the increase in frequency and severity have been attributed to increases in average global air and water temperatures, a result of global warming.

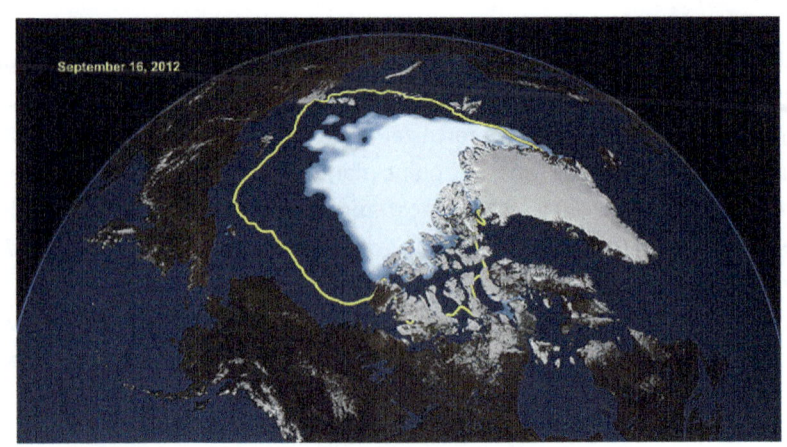

Figure 4.5. As a result of global warming, the amount of ice covering the Arctic pole has never been less. The yellow line depicts the average minimum sea ice between 1979 and 2011. The white area is the amount of ocean covered by ice on September 16, 2012—the lowest amount of sea ice on record to that date.

It is difficult to predict very far into the future but plants have been changing the shape of the planet since they evolved and they continue to shape the globe. We humans have had very little impact on the Earth for most of our existence, which as "civilizations" does not extend much past ten thousand years. However, human populations have exploded over the past few centuries, and the rise of industrialization has given us an unprecedented ability to impact the planet and shape its climate. Over the last 100 years, humanity has impacted this Earth in ways that will last for hundreds—if not thousands—of years.

5

a. the rise of chlorophyll

when earth formed more than 4 billion years ago, it was lifeless. As dust and gases collected, combined, heated, and condensed, the planet we now know was formed along with the rest of the solar system. Earth was initially a volcanic and unpalatable place to live. But eventually, oceans formed and life on Earth began, in what is often referred to as a "chemical soup." During the 1950s, scientists discovered that if they took a bunch of raw materials that would have been present on the Earth when it was young and ran electricity through these materials, simple organic molecules would be formed: This was called the Miller-Urey experiment. Up until this point, no one was quite sure where all the organic molecules came from. These experiments did not create life, but they did create the building blocks for life (specifically, amino acids) and the assumption has been that chance and time would lead to the formation of even more complicated replicating molecules. Eventually, replicating cells would evolve and form more cells that would evolve and turn into things like bacteria, which would then evolve some more and lead to even more complex life forms.

Organic molecules are those molecules that contain carbon. As we know from the first chapter, carbon is an element and necessary for life on Earth. Many of the molecules in your body contain carbon. We are "carbon-based life."

Obviously, there were no power plants or electrical cords billions of years ago, but in order to make new molecules, you need energy. The most likely source of energy was lightning, volcanoes, or ultraviolet radiation. Hence, the Miller-Urey experiment used electricity. If you fast-forward hundreds of millions of years, you get plants, animals, and everything else that walks, crawls, flies, grows, and slimes across the planet. Unfortunately, no one was around to watch this all happen, so much of it is speculation, although it is based on some good experimental science.

The evolution of plants brought a very unique dimension to life on Earth: chlorophyll. Chlorophyll is a clever molecule that acts as the lynchpin of an even cleverer set of chemical reactions: photosynthesis. Chlorophyll is the molecule that makes plants different from just about every other living thing on the planet, it allows them to make their own food. They use sunlight, carbon, water, chlorophyll, and the process of photosynthesis to do this. Before plants evolved, living things had to use chemicals in the environment for their food, generally recognized to be sulphur. This was a pretty limiting way to exist and didn't really allow for much diversity or terribly interesting scenery. So life can exist without photosynthesis, but not a whole lot of life. Plants, however, produce their own food and food for everything else on the planet.

The situation gets a little more complicated, however, when we talk about the origin of plants. Most people think of a plant and something similar to an oak tree comes to mind. However, some of the smallest "plants" are actually microscopic algae and not "true" plants at all. While schoolchildren are often taught that trees and grass produce the air we breathe, at least half of the oxygen comes from phytoplankton in the top few inches of the ocean. And it was the cyanobacteria (also called blue-green algae and a type of phytoplankton) that were some of the first photosynthetic life forms on the planet, going back as far as 3 billion years or so. Cyanobacteria are very small, single-celled organisms that can exist as single cells or small colonies. They can be seen with the naked eye when they group together, but a single cell is microscopic. Cyanobacteria still exist today and ultimately gave rise to the plants we see all around us, but are often dwarfed or ignored in the general discussion about plants.

Figure 5.1. Cyanobacteria (or blue-green algae) are found everywhere throughout the planet and produce a substantial amount of the oxygen on Earth.

Chlorophyll is an old molecule (more than 3 billion years old). But when chlorophyll evolved, the world changed. Until photosynthesis began, the sources of food and energy on Earth were limited. Because photosynthesis requires sunlight as an energy source (and sunlight is renewable and practically inexhaustible), life that could photosynthesize had an unlimited source of energy. With this new source of energy, life could spread. And as primitive photosynthesizing organisms spread, they would serve as a food source for other organisms. Chlorophyll is critically important for another reason, however, and that is oxygen. Whenever photosynthesis is working, oxygen (O_2) is being produced as a waste product. Until the evolution of chlorophyll, there was very little oxygen on Earth, certainly not enough to allow oxygen-dependent life to evolve. As millions upon millions of tiny photosynthetic life forms began to churn out food and spread, oxygen seeped into the atmosphere.

Today, the Earth's atmosphere is made up of about 21% oxygen and 78% nitrogen, with carbon dioxide and other elements making up the rest. When the Earth was young, the composition of the atmosphere was very different. Initially it would have contained large amounts of helium, hydrogen, methane, carbon dioxide, sulfur dioxide and other gases that all would have been toxic to oxygen-breathing life.

We breathe oxygen. All animals require oxygen. Over the course of evolution, we have become dependent upon oxygen for our survival. Had chlorophyll not evolved, we would not have evolved either. The presence of massive amounts of oxygen in the atmosphere not only allowed for the development of oxygen-breathing organisms, but it also demanded it. In most systems, natural or artificial, there are often opposite and opposing forces. When photosynthesis evolved, its opposite also evolved: respiration. When plants produce food from sunlight, carbon dioxide, and water, they evolve oxygen. When we consume that food, we do it by the process of respiration and use oxygen. The process of respiration takes plant food (sugars) and combines it with oxygen to release water, energy, and carbon dioxide. Everything that plants produce, we deconstruct and take apart.

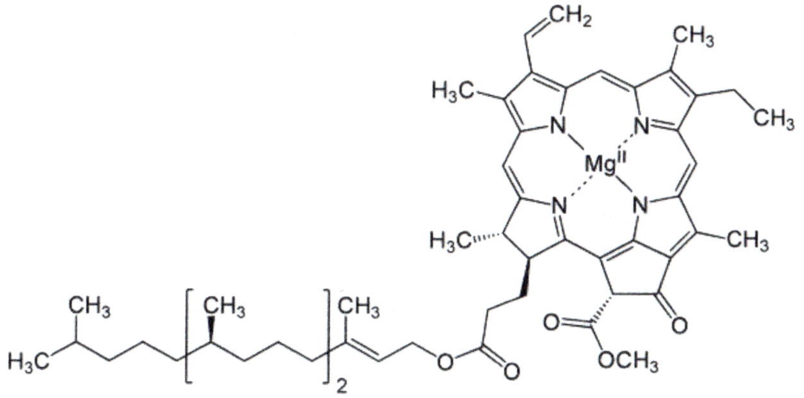

Figure 5.2. Chlorophyll is the molecule responsible for the explosion of life on Earth.

In this way, the energy of the sun is transmitted through the plants and their chlorophyll into every other living thing on Earth. The sun powers us and our Earth—we are of the sun.

Two other things happened when plants started making oxygen. While oxygen is critical for respiration and the primary mechanism by which we extract energy from food, oxygen is also very toxic to many of the organisms that lived on the planet before chlorophyll evolved. Many of those early organisms relied on the process of fermentation, much like when squeezed grape juice is fermented into wine. This process occurs when organic molecules are broken down without oxygen. Many of the organisms that ferment cannot survive in an oxygen atmosphere. As a result, when photosynthesis evolved, much of the early non-photosynthetic life on Earth died out. The other thing that oxygen did for Earth was to provide a protective layer from UV light in the form of ozone, as previously described. If the same amount of ultraviolet light reached the Earth's surface today as it did 4 billion years ago, there would probably be much less life on the planet, despite all the oxygen. Ultraviolet light damages DNA. DNA (deoxyribonucleic acid) is contained in every living thing and provides the molecular instructions that every cell needs to function. When a cell's DNA is damaged, bad things often happen. In many animals, DNA degradation results in cancers. While larger organisms have a little better tolerance (especially if they have hair to block the UV light), bacteria and all those primitive organisms that would have been present when life emerged would have been very susceptible to UV light and may never have survived long or evolved.

Plants require chlorophyll to undertake photosynthesis. It turns out that there are a number of chlorophyll molecules, all similar but with slight differences, and found in plants, cyanobacteria, or both. The molecule itself is enormous, containing 55 carbon atoms, 72 hydrogen atoms, 5 oxygen atoms, 4 nitrogen atoms, and 1 magnesium atom squished in the middle of the molecule. Not only does it have a lot of atoms, but it is very heavy, with a molecular weight of approximately 900. In the context of other molecules, water has an m.w. (the abbreviation for molecular weight) of 18, trinitrotoluene (TNT) has an m.w. of 227, and sucrose has an m.w. of 342.

So we know that chlorophyll is big and complicated, but what does chlorophyll actually do? In the simplest terms, chlorophyll is a pigment. Pigments are all around us. Little kids color with crayons dyed with pigments, our clothes are colored with pigments, and our walls are painted with them. If you asked someone what a pigment is, they would tell you that a pigment is the color of something. But pigments are a little more complicated. They work by absorbing color. When a pigment absorbs a specific set of colors, the wavelengths of light they do not absorb reflect back. Consequently, plants are green because the chlorophyll they contain absorbs light that is red and blue in color, but not green. Plants may be green, but they don't use green light at all! This fact is very important for horticulturalists. If plants are grown indoors, special types of lights must be used that produce red and blue light—the colors that plants require. Without these lights, plants can get spindly and weak. Most light bulbs produce some amount of all the different colors, but plants will not grow very well if they do not get enough of the right colors. When we turn on the lights, we can't see these wavelengths until they get reflected off of something, but they are there.

Energy travels in waves. Different waves have different lengths, and all of these waves exist on the electromagnetic spectrum. The length of a wave determines what kind it will be, what material it can pass through, whether we can see it or not, and whether it is dangerous. Radio waves are very long and don't produce dangerous radiation. X-rays and gamma rays are very, very short and contain a lot of energy. Too much exposure to these can cause damage to tissues and possibly cancer.

Horticulturalists can also get plants to grow better by reflecting red (and even sometimes blue) light back at plants. If a plant is grown in the field, on a red tarp, it will grow more vigorously than a plant grown on a white tarp or no tarp at all because the red light is being bounced back up into the plants' leaves.

Chlorophyll is more than just a pigment, however: It is also an antenna for light. Although all pigments can absorb light, few of them can convert it into anything else. Chlorophyll is an exception. When we listen

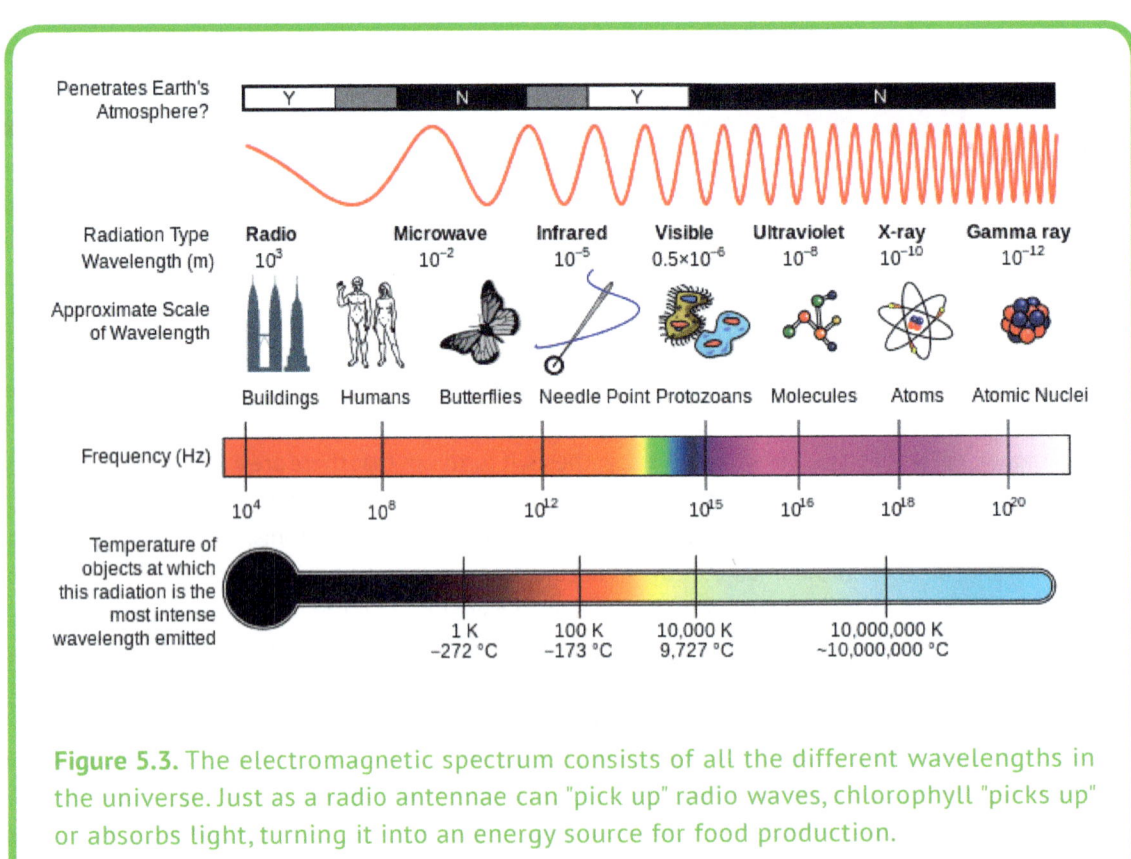

Figure 5.3. The electromagnetic spectrum consists of all the different wavelengths in the universe. Just as a radio antennae can "pick up" radio waves, chlorophyll "picks up" or absorbs light, turning it into an energy source for food production.

to the radio, we are listening to sounds created in a studio, converted into radio waves, and then received by our radio antenna and converted back into sound. While the sound is traveling as radio waves, it is in the form of electromagnetic radiation. Electromagnetic radiation takes many forms, including radio waves, X-rays, microwaves, ultraviolet waves, and visible light. Although they are all part of the electromagnetic spectrum, they all have different wavelengths and do different things. Visible light is a type of electromagnetic radiation, and chlorophyll is one of the molecules that can absorb it and convert that light energy into a different type of energy.

b. how plants make our food

The process of photosynthesis is very complicated, but obviously extremely important to our own survival. Photosynthesis is actually comprised of multiple, smaller steps that work together. It can be divided into what are called "light reactions" and "dark reactions." The light reactions require light; the dark reactions do not.

When red and blue light hit chlorophyll, the energy is received in the form of photons. Photons are the way we describe small and discrete amounts of light. Despite the fact that light usually acts like a wave, it acts like a particle when absorbed by chlorophyll. When photons are absorbed by chlorophyll, the energy that is absorbed causes the "movement" of electrons. In the process, the energy of the sun is converted into something physical, something made of matter. The electrons are moving in a couple of different ways. First, they start to physically move when they absorb this energy, but they also increase in energy potential (they are "moved up" in energy). Electrons don't really appreciate being moved up to a higher energy state. They would much rather not have this energy and will try to lose it once they have it. Once these electrons have this energy, however, the plant can utilize the energy the electrons are happy to get rid of. From these energized electrons, the rest of the process of photosynthesis will proceed.

It should be noted that plants don't pull electrons out of thin air. When chlorophyll uses electrons, they have to come from a source. That source is water. The chlorophyll molecule is constantly extracting electrons from water that it can pass into the photosynthetic process. When water and light are present, chlorophyll will separate the water molecules (using a special part of the molecule containing magnesium ions). This is a very important part of the process, as far as animals are concerned, because this is where oxygen comes from. When water (H_2O) loses an electron to chlorophyll, it turns into two different molecules: oxygen and hydrogen. The plant does not have any use for the oxygen, so it lets it go into the atmosphere. Although plants do require oxygen for respiration, the oxygen is not necessary for photosynthesis so it is lost. It does have a use for the hydrogen later on in the process of phytosynthesis. There are a number of ways this can be done by people, the easiest is to run electricity through water. This process, known as electrolysis, will result in the production of hydrogen gas and oxygen, but requires electrical energy—whereas plants can split water using just the sun's energy.

Once chlorophyll has energized electrons, it sends them on their way. These electrons are passed from different molecules inside the plant along what is called an electron transport chain. As the electrons are moved from molecule to molecule, different events are triggered. For the most part, the electrons result in the movement of hydrogen ions (H). These ions are moved across a membrane and create a gradient that drives further reactions and produces ATP and NADPH. ATP (adenosine triphosphate) is the fuel that powers every cell on Earth. Just like a car runs on gasoline, cells run on ATP. If chlorophyll is the most important molecule on Earth, ATP is surely the second. When an ATP molecule is produced, energy is stored in its chemical bonds.

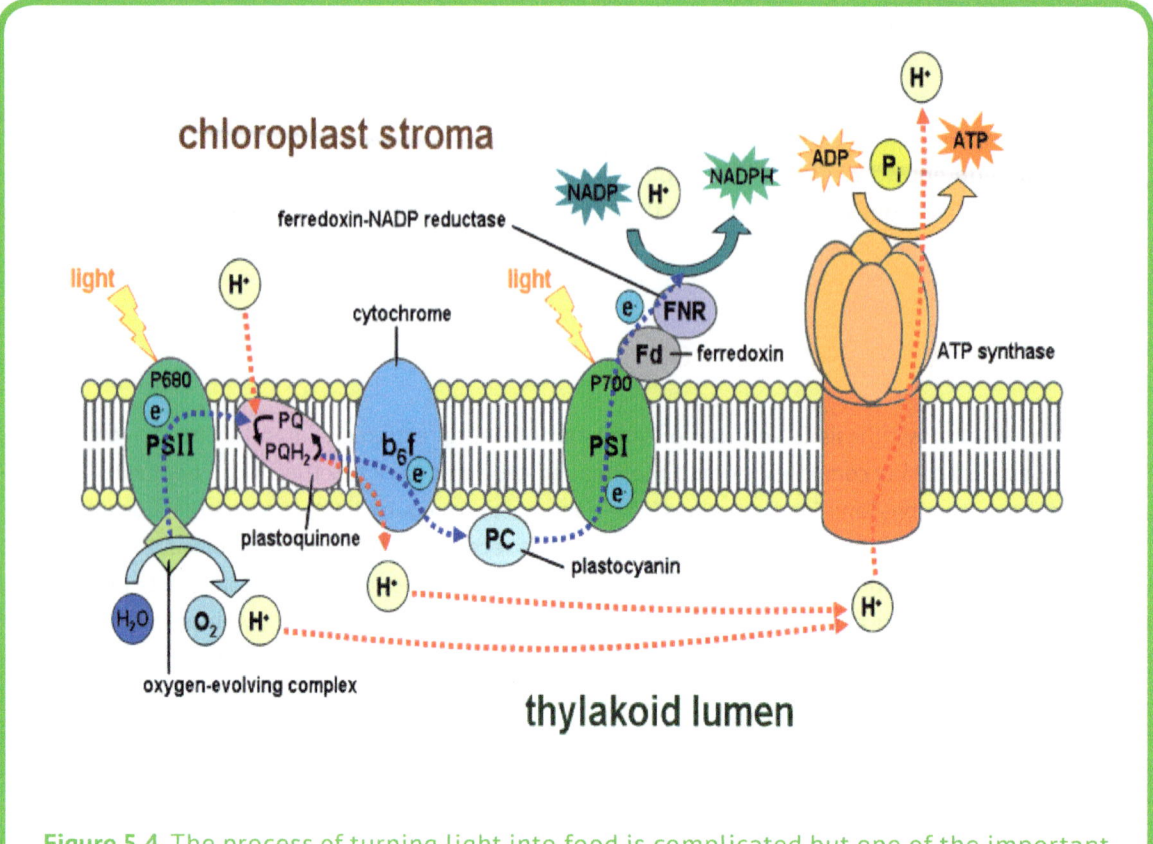

Figure 5.4. The process of turning light into food is complicated but one of the important steps is the movement of electrons through membranes to drive the movements of ions. This is called an electron transport chain.

When ATP is broken down to do work, that energy is released. NADPH (nicotinamide adenine dinucleotide phosphate) is another important molecule because it helps drive the dark reactions. Without this molecule, sugars cannot be produced. Once ATP and NADPH molecules are present, the dark reaction (known as the Calvin cycle) can take place. Plants will take carbon dioxide from the environment and incorporate it into new molecules with the energy provided by ATP and NADPH. The end result is glucose, a sugar that can then be broken down in the future to provide even more ATP, which can be used to power both our cells and the plants' cells. The process is very complicated, much more so than briefly described here. The main point is that chlorophyll is the first step in a long chain of molecular events that start out with the sun's energy and result in the production of sugars. And those sugars are what fuel not only the plants, but all the people too.

c. where it all takes place

Most of the life on Earth takes place in cells. Most organisms are composed of anywhere from a single cell to millions of cells. Cells are where much of the chemistry takes place in life. Individual cells are microscopic: They are too small to be seen with your eyes. Cells are extremely complex and are composed of many different

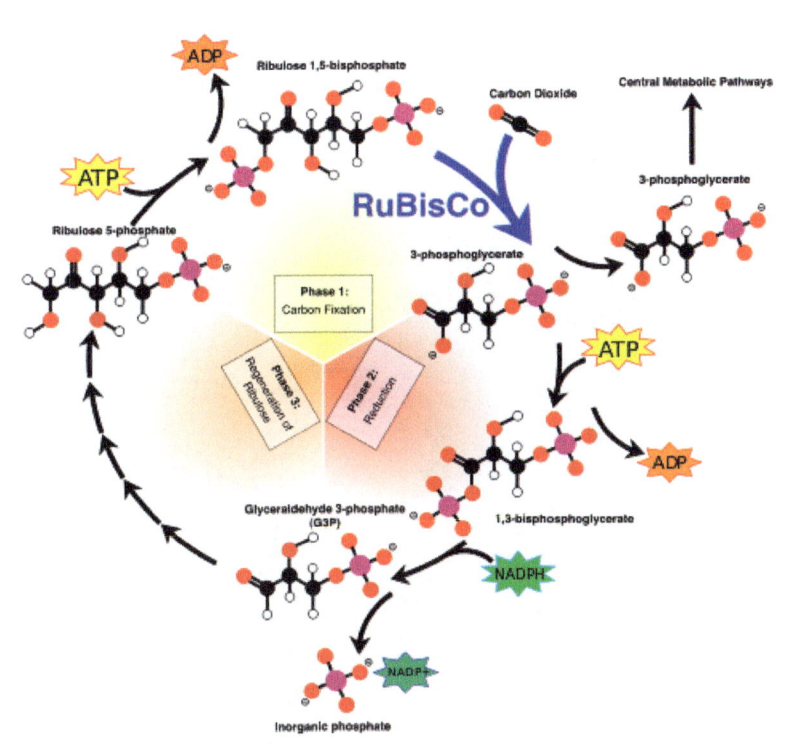

Figure 5.5. The Calvin cycle is the part of the photosynthetic process that does not require light. In this process, carbon dioxide is incorporated into existing molecules (or fixed) to produce 3-phosphoglycerate which can then be further processed, ultimately into sugars.

parts that allow the whole organism to function. Because cells are so complicated, it would be impractical to discuss their working in extensive detail, but a basic understanding of how cells work is important to understand how plants function.

Starting from the outside and working inward, plant cells are surrounded by a cell wall. The wall is just like any other wall—it is a rigid structure that helps provide support and structure. When a bunch of these plant cells are attached to each other, they form a larger solid, rigid structure. People are also made of cells, but one (of many) major difference between plant and animal cells is that animal cells do not have walls. Animal cells are not rigid but extremely flexible. Inside the cell walls and closely attached to it is a cell membrane. The cell membrane is analogous to a balloon. It is very flexible, and it prevents stuff from leaking out and keeps other materials from moving in. Animal cells do have

The primary building block of the plant cell wall is cellulose. Cellulose is a very long molecule made up of thousands of smaller glucose sugar molecules. When these molecules are separate, they look like typical powdered sugar. When they combine in long cellulose chains they look very different, they become strong and rigid cell walls. Most animals cannot degrade cellulose but a special group of animals, the ruminants like cows, contain bacteria that break the cellulose down and release the sugar for animal use.

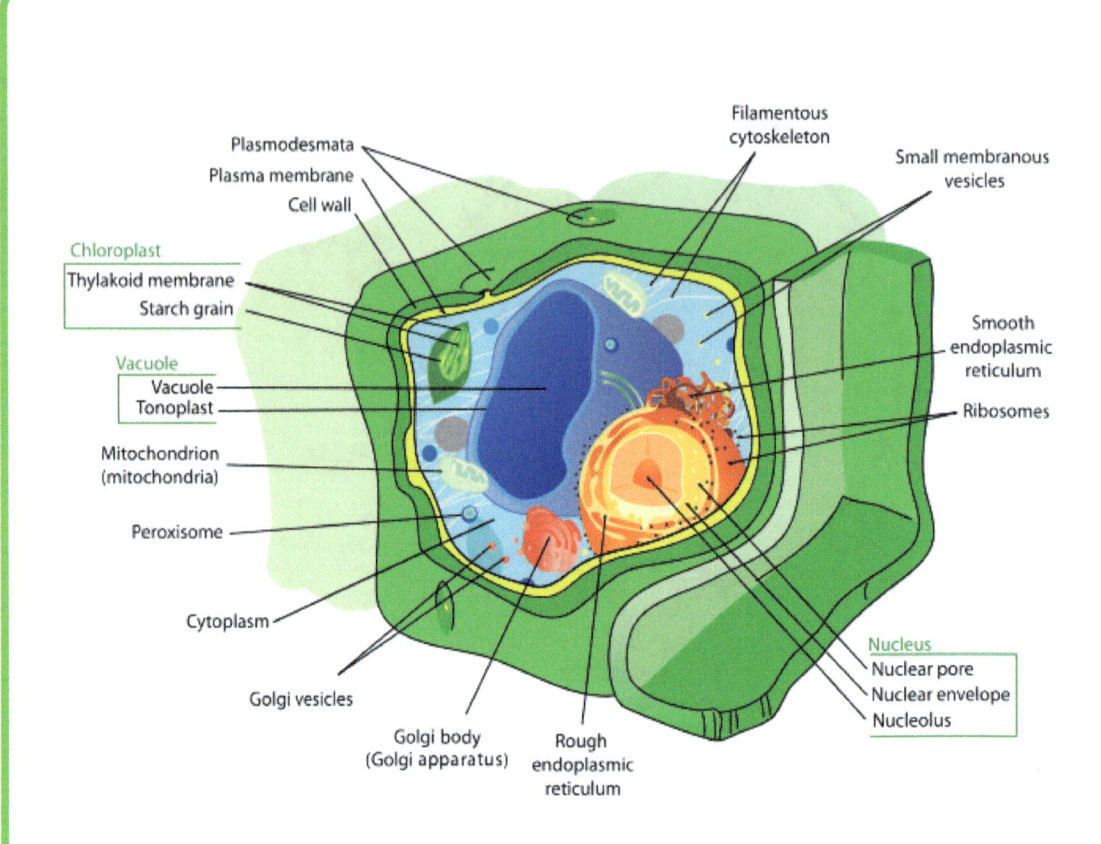

Figure 5.6. Many different parts make up a cell. While animals and plants share many of these parts, unlike animal cells, plant cells also contain chloroplasts for photosynthesis and cell walls for structure.

a cell membrane. Inside the membrane is the place where most of the cellular chemistry occurs. This area is called the cytoplasm. The cytoplasm is a very dense mixture of all kinds of molecules, from things as simple as water to things as complicated as polymerase enzymes. The cytoplasm is where chemical reactions take place, molecules are produced, and other molecules are broken down. The cytoplasm is similar to a the floor of a chemical factory and we can think of it as exactly that, the place where most of the cellular activities take place.

Floating around inside the cytoplasm are a number of smaller parts called organelles. These organelles serve very important and very specific roles. In general, they act as mini-factories. The two major organelles are mitochondria and chloroplasts. Plants have both of these organelles but animals only have mitochondria, with the exception of some Protists which are neither plant nor animal. Mitochondria are the energy-producing part of a call. When a cell absorbs a sugar molecule (sugars are where cells get most of their energy from), it needs to extract the energy out of that sugar so that it can do work. The part of the cell that is responsible for that job are the mitochondria. Mitochondria take sugars, extract the energy out of those sugars, and transfer that energy into ATP. When energy is required in another part of the plant or animal, it comes from the energy stored in ATP. The other major organelle is the chloroplast. Chloroplasts are where photosynthesis takes place and where the chlorophyll is located. Inside the chloroplasts are specialized structures called thylakoids, stacked like pancakes, where the photosynthesis reaction occurs. Inside a plant cell, the chloroplasts take sunlight and convert it into sugars (as described in the previous section). The mitochondria convert those sugars into ATP, which is then used to run the cell.

There are many other parts to a cell besides those previously described, and they all play an important role in cellular function. The nucleus (not the same nucleus in an atom!) is a spherical membrane that looks like a ball and contains all of the DNA which runs the cell. DNA (deoxyribonucleic acid) is composed of extremely long molecules that contain the instructions that run the cell. Because of this, the nucleus is often called the brains of the cell. Cells also contain ribosomes, which make proteins and the Golgi apparatus that processes proteins and a myriad of other parts that contribute to the function of the cell and the plant it is a part of. Cells are extremely complex molecular factories. We have only touched on a small part of what makes them function but what is known about their biology and biochemistry could fit many volumes. And there is much we still don't know about the function of cells and much we are currently learning.

chlorophyll: the most important molecule on the planet

people, plants, and the planet

6

a. evolution is how we got here

the cyanobacteria were the first "plants". They evolved 3 billion years ago and provided a foothold for the plant kingdom, even though they were not really plants. Plants as we know them today evolved from the cyanobacteria. Evolution is the way in which new life develops on Earth. Once life comes into being (as described previously), it starts to change. Evolution requires two different processes or mechanisms. The first mechanism that drives evolution is mutation. Over time, the genetic code (DNA) of an organism will accumulate errors. Many of these errors are fatal—they actually kill the organism in which they occurred or they kill that organism's offspring. However, sometimes a mutation will make an organism or its offspring somehow better or stronger. Being better or stronger takes many different forms. A tree that grows taller than all the rest may be better because it has access to more sunlight. A fish with a more aerodynamic tail may swim faster and avoid predators. Useful mutations make a plant or animal more "fit"; this is where the term "survival of the fittest" comes from. But the height of a tree may also be a disadvantage. Tall trees rarely survive well at high altitudes or in storm-prone areas. Too much height in these areas may make a tree more susceptible to wind damage or damage caused by the weight of heavy snow. The definition of "fitness" depends entirely on your environment. And new adaptation is an adaptation to a specific environment and may provide no benefit or a disadvantage in a different environment.

Mutations in DNA that confer new adaptations are not particularly useful unless a parent can pass them on to their children. When they can, these new traits make the children, the grandchildren, and the great grandchildren more fit and more likely to thrive. Over time, these descendants will become more numerous. This is the second mechanism of evolution, natural selection. Although many people credit Darwin with describing evolution, he actually described the process of natural selection. Evolution was not a new concept when Darwin started to investigate it but no one knew how it could possibly work. Natural selection (the *process* where the strongest or "fittest" and most suited organisms conquer the weaker organisms) is how evolution works. And after many years of diligent work, Darwin demonstrated evolution through natural selection.

Evolution works very slowly, and it works on every aspect of a plant, animal, fungus, bacteria, or virus. As mentioned before, while some mutations turn out to increase fitness, others that occur at the same time may decrease fitness. After millions of years, where there was just one species of animal, there may now be 3 or 5 or 100. There are actually some additional processes that can affect evolution (and which we will not discuss here), but mutation and natural selection account for the most significant factors that allow for changes that result in new populations and new species over time.

a plant genealogy

When the cyanobacteria (the microscopic blue-green algae) first evolved, they were very simple. Cyanobacteria still exist today and are still very simple. But as mutation and natural selection took place, more advanced forms of cyanobacteria developed. Cyanobacteria eventually evolved into actual green-algae, which contained more complicated cellular machinery and utilized a more complicated method of reproduction. Approximately 500 million years ago, these algae evolved into simple, multicellular plants that could live on land, but were generally relegated to the wettest of locations. From these, the liverworts and mosses evolved, plants that also still exist today. But while mosses are still very common plants (also known as Bryophytes), they are still not very competitive or very advanced. Bryophytes lack a critical adaptation that limits their success: They do not have vascular tissues. The term vascular refers to tubes that move liquid. The vascular system in an animal moves blood. The vascular system in a plant moves water. And until plants had an organized system for moving water, they could only grow in wet places. Plants generally stayed relatively small because they could not move water very far upward from the soil. The first vascular plants to evolve from the Bryophytes were still small, but the vascular system allowed the move further inland. It cannot be understated that water is critical for life and that life generally evolves in (and remains near) water sources. Water could put up a very good fight for the title of the most important molecule on Earth! Once vascular systems developed, plants evolved roots (for collecting water from the ground and anchoring), leaves (for absorbing sunlight), wood (for maintaining rigidity and increasing height), and seeds (for dispersing new plants over great distances). Within 100 million years, plants had spread rapidly across the landmasses of the Earth, where they formed huge forests of ferns and trees.

The last major evolutionary event we'll mention is the development of flowers. Before plants had flowers, they spread primarily using spores and sometimes seeds. Flowers are another way to produce seeds but flowers are unique and have only been around for the last 130 million years. The main advantage that a flower bestows upon a plant is the ability to use insects to aid in plant reproduction. Seeds are a reproductive structure. In order to make a seed, plants needed to

Figure 6.1. Although Charles Darwin is famous for his work on evolution, most people mistakenly think he proposed the theory. In fact, Darwin's contribution to biology was to figure out how evolution actually worked, not to propose the theory itself.

Not all plants or animals that evolved over the life of the planet still exist today. As climate changed or food sources became scarce or as predators killed them, many species went extinct. While the dinosaurs are most conspicuously absent from the planet these days, many thousands of plant and animal species are also long gone. And species continue to go extinct even today—often with the unfortunate aid of human populations.

combine the male and female sexual organs. Seeds lead to new genetic combinations, just like a child is a new genetic combination of his or her parents. New genetic combinations are important because they aid the process of natural selection and evolution. However, until flowers developed, producing seeds was not particularly efficient. Flowers allow plants to produce seeds more reliably. As insects evolved, flowers on plants evolved. Flowers are visually appealing to insects, particularly bees. When they land on a flower, they pick up pollen. Pollen is the male reproductive part. When the insect goes to another flower, it deposits the pollen in the other flower, where the male pollen unites with the female part. The results are seeds. The process of moving pollen is called pollination. Most plants that produce flowers use the insects as their tool for producing new seeds. Once the flower evolved, flowering plants exploded across the globe. They were able to produce massive quantities of seeds and these seeds spread. The flowering plants are called the Angiosperms and are the most successful group of plants on Earth. From food crops to textiles to construction materials, the Angiosperms provide us with most of the raw materials we use today.

Not all flowers use insects. Grass plants are often pollinated by wind. Because they do not use insects, their flowers are green and brown and unappealing to insects. Other plants don't use either insects or wind and have learned to pollinate their flowers without any other help in the process of self-pollination.

b. how to tell the plants apart

When most people look at a plant, the feature they most often see (and remember most easily) are its leaves. Different plants do have different shapes, colors, and sizes of leaves. However, leaves can change. One of the easiest evolutionary changes for a plant to make is to change the shape, size, or color of its leaves. In nature, changes in leaf shape are relatively common. Most of the leaves of red oaks look very similar, if not the same, but there are dozens of different species in the red oak group. Many of these species have similarly shaped leaves, but they are all quite different.

Figure 6.2. Liverworts were some of the first plants to evolve and continue to survive in unique ecological niches to this day.

The term "species" has not yet been discussed very much, but it is an important concept in biology. A species is often defined as a group of related individuals but it is more than that. All people belong to the same species, *Homo sapiens* (more on these Latin names in the next section). A red oak tree belongs to the species *Quercus rubra*. A black oak, however, belongs to the species *Quercus velutina*. A lot of things make a red oak different from a black oak. The most similar thing about the two is leaf shape. But the reason they are different species is that they cannot reproduce together. The definition of a species is not only a group of organisms, but a group that is able to reproduce within that group. This is actually a very simplistic way of defining a species, but it works well with plants in natural settings (less so with things like bacteria). In artificial settings, we can make different species reproduce with each other that would not do so in nature, but that is an exception to our definition.

So the question remains, How do we tell different plants apart, how do we determine which species a plant belongs to? We could look at other features besides leaves. Perhaps the bark of a tree? The buds at the ends of the branches? Size and shape of the plant? Or even the fruit the plant produces? Unfortunately, none of

Figure 6.3. Every plant species is different but the unifying theme among plant families are flowers. Because the flower is a reproductive structure, it changes little over the course of evolutionary history while structures like leaves change dramatically.

these features can give us a conclusive identification by themselves. Even when we combine these different features, we may still be unsure of a plant's identity. The lynchpin, the feature that ties everything else together, however, is the flower (or in a plant without flowers, alternate reproductive features). Although all the parts of the plant are susceptible to evolutionary change, the reproductive structures are the character that changes the least. Angiosperms use their flowers to reproduce. Without their flowers, flowering plants cannot produce seeds. If they cannot produce seeds, they die out. Consequently, it is generally a bad idea for plants to start making changes to their flowers because it may mean that they cannot reproduce with other plants and they will die out, never producing any offspring. This does actually happen from time to time, resulting in an individual plant being unable to reproduce, but it goes unnoticed as the rest of the plants in that species continue to produce offspring.

Because flowers and other reproductive structures change so infrequently, we can use them as a major component in identifying plant families and even plant genera and species. Without much effort, it is very easy to tell what family a plant belongs to just by looking at its flowers. Unfortunately, plants do not produce flowers all the time so this can complicate things. When a plant does not have a flower, we need to combine all of its other features to try to identify it. And there are many plants that are not Angiosperms and never produce flowers. In these cases, all the other parts of a plant become critical in identifying it. Other methods for identifying plants also exist, particularly DNA analysis. Although all people are people, the DNA from one person to another is subtly different. But it would be easy enough to look at the DNA of a modern human (*Homo sapiens*) and tell that is not from an ancient species of man, such as the Neanderthals (*Homo neanderthalensis*). The same can be done with any plant species. The process is relatively straightforward and only takes a few hours in a laboratory, unfortunately the cost of equipment, supplies and the technical expertise to do DNA identification of plants can be very expensive. While it can be very useful to researchers, most horticulturalists and plant biologists identify plants the old fashioned way- by looking at their parts, particularly their flowers!

c. the way biologists talk

Even when we can identify a plant, that plant may have many different names. The same plants often evolve and flourish in many different places simultaneously, and different civilizations will give the same plants different names. This can be extremely confusing, especially to scientists and biologists. Fortunately, there is a way to identify every plant and every living thing on the planet in a way that prevents confusion. This is done using a system called binomial nomenclature. The word binomial means "two names." Nomenclature refers to a naming system. Biologists and other scientists give every living thing a two-part name to separate it from every other living thing.

Using a system of two-part names is certainly clever, but the system is even more universal because every two-part name is in Latin. Scientists from different countries speak different languages. It would often be impossible (or at least extremely confusing) if they were to try and translate a plant's two-part name from one language to another. Many languages do not share the same words and words can have widely different meanings in different languages. Consequently, if everyone uses the same language (Latin) for binomial nomenclature, there is very little confusion. This system works very well. Often, these names are called Latin or scientific names.

The idea for this simple naming system came from Carl Linnaeus, a Swedish botanist who first published the idea in 1753. He then proceeded to name everything he stumbled upon, providing thousands of scientific names for plants and animals. Before Linnaeus's system, Latin was often used in the names of different species, but the names were long and cumbersome. Instead of just two simple words to name a plant, the name might contain half a dozen different Latin words. And those earlier names weren't always grouped together. As a consequence, two plants that were very closely related might have very different names. The other strength to Linnaeus's binomial nomenclature is that all living things are also placed in groups with other things that they are related to. As taxonomists learn about new connections and new relationships, names can change

Figure 6.4. Carl Linnaeus, a Swedish botanist, first published the idea for the binomial nomenclature system we still use today for naming life in 1753.

and even if they don't change, the groupings can. Linnaeus's system was short, simple, and eventually received universal acceptance.

We've already mentioned some scientific names in this chapter: *Homo sapiens*, *Quercus rubra*, and *Quercus velutina*. These names are composed of two parts, as are all scientific names. The first part is the genus: *Homo* and *Quercus*. These parts do not refer to a single species, but rather a group of closely related species. Anything in the genus *Quercus* is an oak tree, but there are lots of different oak trees. The second part (*sapiens*, *rubra*, and *velutina*) is the specific epithet, or species name. By itself, the second part of the name does not mean anything. When it is combined with the genus name, however, then you have a species. *Quercus* means the oak trees. *Rubra* is just the Latin word for red. *Quercus rubra* is a red oak.

This system works for all living things. Everyone has seen different types of cats. Different cats have different scientific names. The house cat is *Felis catus*. The bobcat is *Lynx rufus*. The lion is *Panthera leo*. The leopard is *Panthera pardus*. What you should notice is that all the names are different, but lions and leopards share the same genus name. This means that they are very closely related, more closely related to each other than to the house cat or the bobcat. And the bobcat is not very closely related to the house cat, even though they look similar and are a similar size. And we know this because they are different genera: *Felis* (housecat) vs *Lynx* (bobcat). And there are actual lynxes, such as the Canada Lynx *(Lynx canadensis)* which is closely related to the bobcat but different enough to be a separate species.

It should be noted that there are some rules governing scientific names. First, scientific names should be set apart from the rest of the text when writing them. Usually this is done with italics or with an underline. In addition, only the genus name is capitalized; the specific epithet is always in lower case letters. There are even more rules, many governing how to name a new species, but they can get very complicated. There is a last part to a scientific name that has not yet been discussed: the namer. The person who names a new species has their last name attached to that species name. This part is often skipped in common practice, but it is part of the fully recognized scientific name. Because Linnaeus named so many things, his last name is often attached and abbreviated "L." For example, *Poa annua* L. is the full scientific name for annual bluegrass. When another individual has named a plant or animal (or any other living thing, for that matter) their name is attached, as in the case of *Paspalum dilatatum* Poir., the scientific name for Dallisgrass, named by Jean Louis Marie Poiret. And sometimes, multiple people have named the exact same species. In this case, the species is given the name attached to it by the first person proven to document the species, assuming they placed it in the proper genus. If not, attribution of a name can get very complicated! Scientific names are not always permanent. As new relationships between genera are discovered, names can change.

d. the plant families

Just as every person has a family and is descended from parents and grandparents and great-grandparents, every plant, animal, fungus, or other living create is descended through millions of years of evolution from a progenitor species and belongs in its own closely related group. As we have already mentioned, plant evolution started with cyanobacteria, which evolved to become more complex and eventually gave rise to the flowering plants, or Angiosperms. Over evolutionary time, however, many different Angiosperms have evolved. Some are closely related to each other and some are more distant. When biologists talk about life on the planet, they classify it based on how it evolved and how closely one plant or animal is related to another plant or animal.

If we were to consider all the life on Earth, we can break it down into a number of very large categories. One way to divide life is to simply place it into five categories, or kingdoms: Plants, Animals, Fungi, Bacteria, and Protists (which are generally a group of microscopic organisms that just do not belong to the first four groups). Another way we can divide up life is into three domains: archaea (extreme bacteria, these bacteria are the oldest and live in harsh environments), bacteria, and eukaryotes (everything else). And then there are some versions that blend these two and arrive at various configurations with six groups. Some scientists believe there should be even more kingdoms, dozens of kingdoms, but five to six is usually where most people settle. The problem with placing life into categories is that life did not evolve in distinct categories. Life is messy. And it has been messy for hundreds of millions of years. Life involves many random things and many millions of years of trial and error. Science, on the other hand, tries to be very neat. Science does its best to figure out how to classify life but sometimes life can defy classification and scientists often disagree on the best approaches, based on their own biases and experience.

Once an organism is in a kingdom, it will also belong to a subdivision, usually called a phylum (or a division, in plants), followed by a class (or subdivision), followed by an order, followed by a family, and finally followed by the genus and species. But determining evolutionary relationships can be complicated and confusing. It can be messy. As a result, biologists will disagree about the exact phylogenetic tree (or branch on the "tree of life"- similar to a family tree) that a plant or animal belongs in. These phylogenetic trees (from kingdom to species) may include additional levels that are only appropriate for certain groups of plants, animals, or other organisms.

When talking about plants, families are the next most important level of classification after genus and species. Those plants within the same family often share many of the same characteristics and can easily be grouped together. They will have similar flowers, they often look the same, and they will often share the same uses. There are currently about 400 different recognized plant families. Plant families always end in the letters "aceae," making it easy to tell a family from an order (which ends in "ales") or any other classification.

Some of the common families are Poaceae (grasses), Fabacaea (peas), Solanaceae (potatoes and tomatoes), Rosaceae (roses, apples, pears), Magnoliaceae (magnolias), and many others. Each of these families has many different species that share common traits. Most people would not think that a rose and an apple are closely related, but they both share a common lineage—along with strawberries, pears, almonds, raspberries, and still others. Millions of years of evolution have separated roses from apples, but they still reside in the same family, despite their apparent differences.

7

a. seeds are plants on the move

plants typically evolve in one location, called a Center of Diversity or a Center of Origin. These will be detailed thoroughly in Chapter 12, but from that location, plants spread to other environments. All creatures, whether they are plants or animals, have an innate need to procreate and spread. Life on Earth creates more life. And it is more than a need; it is a function of existence. All life—viruses, bacteria, fungi, insects, fish, birds, cats, dogs, and people—procreates in some manner. And as more life is created, it needs room to thrive. Plants, unlike living things that move freely, are constrained by their inability to travel. Plants cannot "pull up their roots" and travel. As a consequence, they need a mechanism to move beyond their boundaries. Over the course of evolutionary history, plants could have developed any number of means for moving themselves from place to place. The method that did evolve was to employ their seeds. Other plant parts can be dispersed to produce new individuals and found new populations, but the seed is the most important resource to enable plant species to travel beyond their current boundaries.

Seeds cannot travel on their own, however. Wind, water, animals, and even gravity assist with the distribution of seeds. Small seeds, like those from orchids, petunias, and many grasses, may be easily spread via wind. In addition, their very small size keeps these seeds from being eaten by animals. Unfortunately, a small seed may have a low survival rate because of the very small amount of nutritive resources stored in the seed. Plants growing near water and whose seeds have buoyancy may be swept away by ocean currents and travel vast distances from the parent plant. Coconuts (*Cocos nucifera*) are a prime example of this method of species' movement. Coconut fruit are very large (the coconut you eat is the fruit with a seed contained within it) but they float. Coconuts can be produced on one island, fall into the ocean, and wash ashore hundreds of miles away, where they germinate and produce new coconut palms.

Animals may carry plant seeds inside their digestive system after the animal consumes the seeds, usually while they are encased in the mature fruit that attracted the animal to eat the seeds along with the fruit. The seeds have coats that are strong enough to withstand complete dissolution by the acids in the animals' digestive tract. After passing through the animals' digestive system, the seed coats are usually etched enough to allow seed germination once a suitable growing environment occurs. Birds are excellent at transporting seeds because they do not chew the seeds. When a bird lands on a raspberry bush, it is attracted to the fruit the raspberry produces. The raspberry wants to disperse its seeds so it produces sweet and sugary raspberry fruit, with seeds contained inside. Birds eat the fruit, fly away, and spread the seeds far and wide as they defecate. However, when a mammal eats the raspberry, it may crunch the seeds in its teeth. Seeds that get chewed will be destroyed and cannot spread. Consequently, many plants produce seeds in fruit, optimized for bird consumption and dispersal. Oak trees (*Quercus* spp.) are exactly the opposite. Birds don't have much use

the spread of the plants

for the acorn that the oak tree produces, but squirrels do. Squirrels eat a lot of acorns but in the process they also bury many acorns and lose track of them. Those lost acorns germinate and become new oak trees. Oak trees and squirrels work together so that both of them survive and proliferate. In fact, when oak trees have a bad year and don't produce enough acorns, the squirrel populations suffer without this staple food source.

Seeds surrounded by protective structures that are sticky or have barbs may become attached to animal furs or feathers for transport to other regions. Birds carry some seeds (like the sticky ones from mistletoe) into trees, where the seeds are rubbed or cleaned off the birds' bodies and become attached to suitable

Figure 7.1. Some seeds utilize their size for protection. In the case of tobacco, the seeds are extremely small—so small that they can hide from herbivores easily in the soil.

niches on the trees. Plants that grow on other plants without their roots becoming anchored in the soil are known as epiphytes. Some epiphytes, like mistletoe, become parasitic to their host plant by penetrating into the host for its food source. Other epiphytes simply use their host for support and make their own food.

The barbed structures on the seed coats of plants like burdock (*Arctium* spp.) enable the seeds to attach to softer surfaces like an animal's fur and people's clothing. Anyone who has walked in a natural habitat where these plants exist has likely been exasperated by the abundance of these seeds clinging to their clothing. It was just such an encounter in 1948 by George de Mestral from Switzerland that led to the development of the ubiquitous fastener known as Velcro.

Plants growing on hilly regions may have their seeds carried down the hills by gravity. All of these means of distributing seeds away from the plants that produced them serve two purposes. Primarily at the individual level, the seedlings have a better chance to become established without competing for resources with the parent plants. When too may seeds germinate in the same place at the same time, they compete for water, sunlight, and nutrients. As a result, the new plants struggle to survive, are susceptible to diseases and drought, and perhaps none of them will survive. Secondarily, at the species level, seedlings that move into new areas may be able to adapt to new habitats different from the native habitat of the parent. If the original habitat becomes unsuitable for a particular plant's survival, the movement of offspring to new habitats may enable the species to avoid extinction.

Seeds come in all different shapes, sizes, and colors. These observable differences serve many purposes. Small seeds fall into cracks and crevices, are hard for animals to eat, and can be produced in the thousands or millions. Large seeds are more noticeable to animals and foragers, but they have more nutrient reserves and may last longer in the environment. However, a plant can produce only a limited number of large seeds, thus

Figure 7.2. When a seed germinates, the immature root (radicle) and leaves (cotyledons) emerge and grow gravitropically.

reducing its chances for survival. Seed color may allow a seed to be camouflaged in one environment and not another. As was mentioned previously, structures on seeds may aid in dispersal. Clearly there is a huge amount of diversity among all the seeds that exist and the uniqueness of each seed type allows it to be successful in the environment it is adapted for.

Seed structure is primarily dependent upon whether is it a monocot or a dicot seed. The terms "monocot" and "dicot" refer to the type of structures inside a seed and are abbreviations of monocotyledon and dicotyledon. The cotyledon is a special leaf produced in the seed. It is called a seed leaf and its primary purpose is to feed the young embryo just after it germinates. Because a newly emerged seed cannot produce much of its own food, the cotyledon is critical in fueling the newly developing plant. A monocot has one seed leaf (mono = one) and a dicot has two seed leaves (di = two). There are additional differences besides just the number, particularly in the fashion that these leaves emerge, but it's not critical for our current discussion.

All seeds will have a protective seed coat. Inside the coat is found the embryo, the most important part of the seed. The embryo is the immature plant. In a dicot plant, the embryo will be attached to the two cotyledons that will provide it with a food source until it begins to photosynthesize. In a monocot plant, the embryo will be attached to only a single small cotyledon, but the rest of the seed will be full of endosperm. Endosperm is similar to the cotyledon in that it provides nutrition for the emerging seed. The endosperm is also important for people because of its use: in wheat and other cereals, the endosperm is the part of the seed that is ground into flour. Endosperm does sometimes occur in dicots, but it is typically very small and quickly gets utilized for the development of the cotyledons. More differences between monocot and dicot plants will be described in the next section.

Most people don't realize it, but seeds are living things. While a seed may not appear to be

Figure 7.3. Sunflowers are dicots. As the seedling emerges, the cotyledons are the last part to exit the seed coat and will only last a few days before they are replaced by true leaves.

doing much, it is actually alive but dormant. Seeds rarely grow when they are first produced. Instead the seed stays in a dormant state, as though it were sleeping, waiting for the appropriate conditions to occur before it germinates. The process of germination is the process by which a seed wakes up and begins to become a new plant. Conditions vary but seeds usually require a minimal level of moisture (they will absorb water through the seed coat), the right temperature (seeds that germinate when temperatures are too cold or to hot may die quickly), and oxygen. When a seed begins to germinate, it will need to respire stored carbohydrates for energy. Respiration requires oxygen and seeds that are growing in a suffocating environment, such as a densely packed soil, will quickly die.

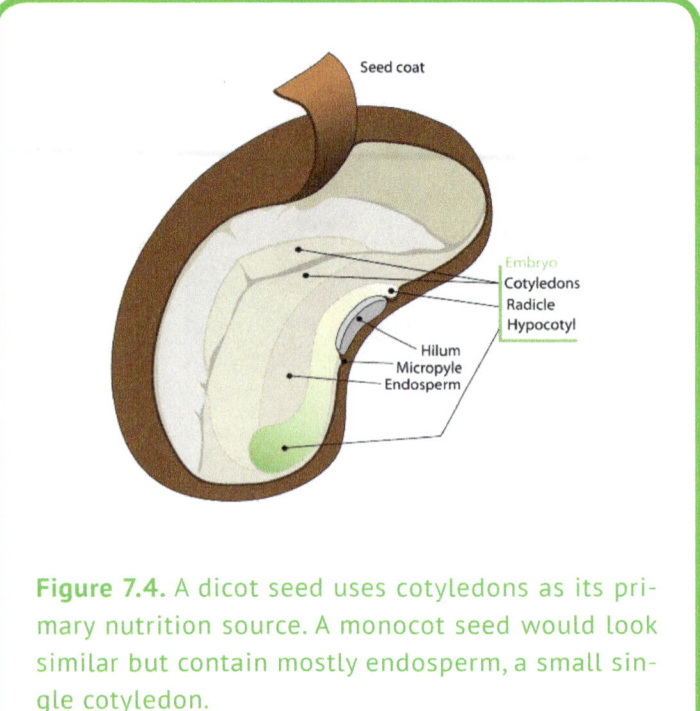

Figure 7.4. A dicot seed uses cotyledons as its primary nutrition source. A monocot seed would look similar but contain mostly endosperm, a small single cotyledon.

A seed that never has the opportunity to germinate may die. Although some seeds can stay alive and dormant for hundreds to thousands of years, most seeds have much shorter life spans.

b. where seeds come from—flowers!

Where do seeds come from? The answer is simple: flowers. But flowers, their structures, evolution, and function can be complicated. First and foremost, it is important to realize that a flower is a sexual structure. Male parts (pollen) of the flower fertilize the female part (ovules) of a flower and an embryo is produced. The seed is the structure that contains the embryo and is a sexually produced offspring. That means that half of the genetic material in the seed comes from the male part and half comes from the female part.

In some species, some plants are male while others are female, but most plants have flowers that contain both male and female parts or a plant may have two different types of flowers, male and female, growing on the same plant. Before fertilization can occur (which is similar to fertilization in humans or any other animal) pollination has to occur. Pollination and fertilization are two different things; pollination leads to fertilization. Pollination is the process of moving the male pollen to the female stigma, which is attached to the ovary and ovules. When a pollen grain lands on a stigma, it grows a germ tube that goes through the stigma and into the style, and reaches the ovule. When that happens, fertilization can occur.

While not all seeds come from flowers, flowering plants are the most prolific on the planet and most seeds are produced by some type of flower. All flowering plants are classified as angiosperms, including plants we do not usually categorize as flowers. This includes many deciduous trees, shrubs, vines, and grasses. The gymnosperms (or conifers) also produce seeds but do not produce flowers. Angiosperms can be further divided into two large categories: the monocot (monocotyledon) and the dicot (dicotyledon). Floral structures of dicotyledonous and monocotyledonous angiosperms are similar in many aspects, while different in others.

It is important to realize that a flower is just a specialized leaf. Leaves evolved before flowers and when the evolution of flowers began, leaves were the raw material that evolved into floral structures.

Figure 7.5. Monocots (*Miscanthus* spp. on the left) have vascular bundles running parallel to each other in leaves. Dicots (*Ribes* spp. on the right) have vascular bundles running in many different directions and forming branching patterns.

Evolution will often modify an existing structure, as opposed to developing a new one from scratch. All complete flowers contain the four basic floral parts: sepals, petals, pistils, and stamens. These four components are attached to a base, known as the receptacle. If one or more of the basic parts is missing, the flower is described as incomplete. Not all parts of the flower are critical for it to be functional and some plants don't produce them all.

Sepals are an excellent example; they are not evident on all flowers, especially monocot grasses, which have unique structures, described later in this section. When present, sepals usually function to protect petals, pistils, and stamens from adverse temperatures and pest damage while the petals, pistils, and stamens are developing during the bud stage. Sepals are typically green, although they may be other hues. Occasionally, the sepals may be fused together. Collectively, all the sepals on a flower are referred to as the calyx.

Petals are often the most distinguishable feature of flowers. They have a range of colors from white to red to dark blue, which may be a solid color or patterned. Some flowers, such as some tulips (*Tulipa* spp.), originally obtained two-toned coloring from mosaic viruses located within the plant. Breeding and selection now account for many species' multiple coloration, including tulips. The color and/or patterns of some flower petals may serve to direct the animal that helps pollinate those flowers. Hummingbirds favor the color red, while blue and purple are preferred by honeybees. The relationship between flowers and animal pollinators is extremely important. Almost 80% of the flowering plants rely on insects and birds for pollination. The relationship is one that has evolved over millions of years and allowed plants to maintain genetic diversity and spread while providing a primary food source for animals. In fact, much of the plant products we consume today depend on insects pollinating flowers that develop into fruit. Without insects, particularly bees, much of the food we buy today in the store would be unavailable.

Flower petal shapes are varied, with some petals fused to each other or even to the sepals. Collectively, all the petals on a flower are referred to as the corolla. If the petals are fused into a tube shape, this is known as a corona. When the petals and sepals are indistinguishable, as a group they may be called tepals.

Animal-pollinated flowers tend to be more prominent than wind-pollinated flowers. Colors, markings, shapes, or scents attract animals essential to pollination of many angiosperms. Petals may have markings that direct insect pollinators toward the female and male reproductive structures of the flower. Conversely, many monocot plants are wind-pollinated and nondescript. That is, because the wind is responsible for moving the pollen to the stigma, grass flowers can be boring. They just need to allow the wind to deposit pollen on a stigma, not attract the attention of an animal to do the job. These flowers may have structures unique to monocot species. Monocots typically have three petals or some multiple of the number three. Dicots often have four or five petals. They may also have multiples of four or five petals.

Figure 7.6. All of the sexual structures of the pear flower are easily seen. Each one of these flowers will develop into pears if they are pollinated and fertilized.

"Pistil" is the collective term for the three female parts of a flower (stigma, style, and ovary). In a typical dicot flower, the pistil is often oar-shaped or shaped like the pestle of the grinding tools mortar and pestle. Flowers may have more than one pistil, which varies by species, which means the flower will also have multiple ovaries and can produce more than just one seed. The stigma is the outermost structure of the pistil. It may be sticky, with a fluid excreted by the flower, to promote adherence of pollen grains. Some stigma surfaces may be rough to help trap pollen grains during the fertilization process. Ovules (unfertilized eggs) develop in the ovary, which is usually located toward the base of a flower.

The style connects the ovary and the stigma. Wind-pollinated flowers often have elongated styles to elevate the stigmas, exposing them to pollen passing by the flower in the wind. Styles are almost non-existent in other flowers. In many monocot flowers, pollen is trapped in two feathery stigmas attached to the ovary by a short or nearly absent style.

The male portion of the flower is the stamen, which has two parts: the anther and filament. The anther contains the pollen grains and the filament connects the anther to the receptacle. When the pollen is mature, the anther splits open to release the pollen. Sometimes, the pollen is forcefully ejected by the rapid opening of the anther.

Another way flowers are classified is by the presence or absence of functional pistils and stamens. If both pistils and stamens are functional within a flower, the flower is considered "perfect" and has *both* male and female parts. If only the pistil or the stamen is functional, the flower is classified as "imperfect" and thus has *either* male or female parts. Female flowers are known as pistillate flowers, while male flowers are staminate. Species with separate female and male flowers on the same plant are monoecious. Corn (*Zea mays*) is a well-known example of a monoecious plant species. The ears are produced from female flowers on the plant's

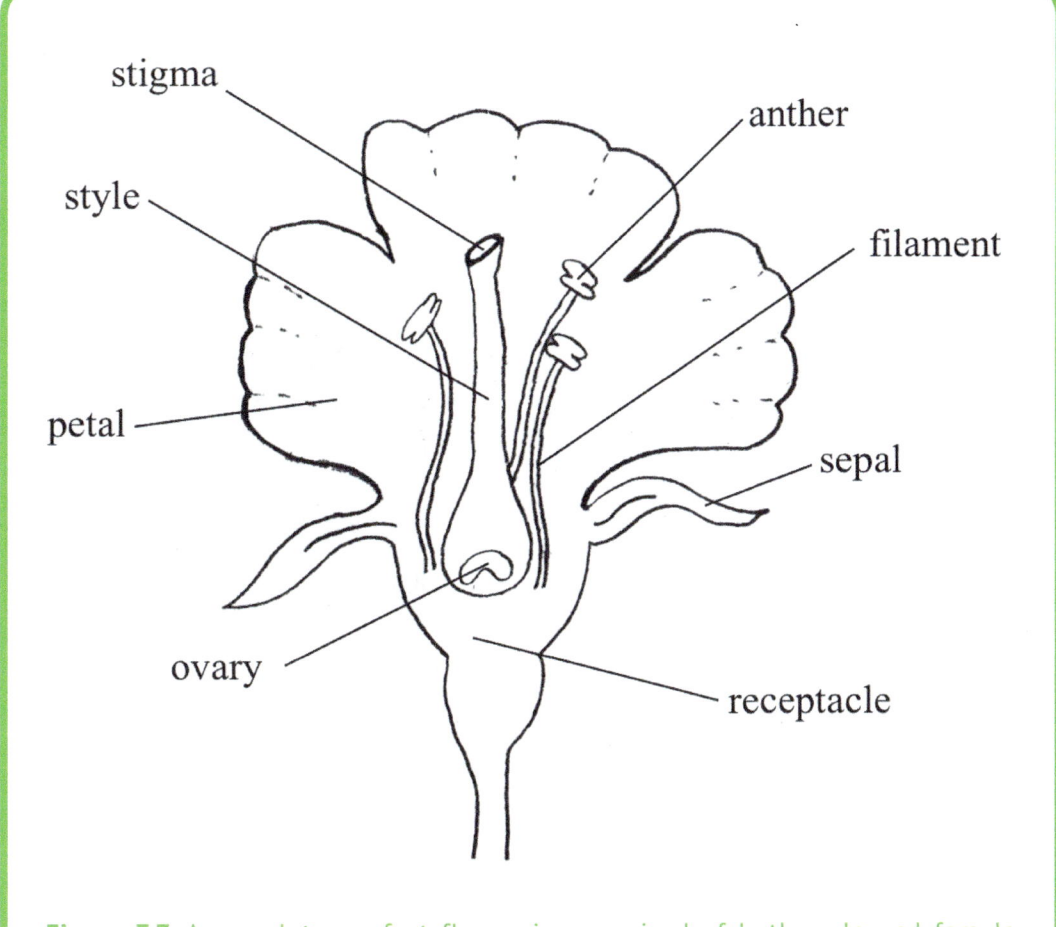

Figure 7.7. A complete, perfect flower is comprised of both male and female sexual parts and all of the floral components like sepals and petals.

stem. Tassels atop corn plants are the male flowers. When male and female flowers are found on separate plants, the species is dioecious. Examples of dioecious species include a native prairie grass known as buffalograss (*Buchloe dactyloides*), bananas (*Musa* spp.), holly (*Ilex* spp.), and asparagus (*Asparagus officinalis*). One plant of each sex is necessary for seeds to be produced on the female plants. The male plants produce only pollen from their flowers and never fruit. Sometimes this can be problematic. Holly plants are grown as ornamental and produce bright red berries that are considered attractive by some people and a nuisance by other people. If only male or female plants are used in a location, no berries will be produced. Both male and female plants must be present to get berries but male plants will never produce berries, only female plants.

On many monocot flowers, sepals and petals are replaced by structures known as glumes, lemmas, and paleas. These flowers occur on many of our grains, including barley, oats, and wheat. What is normally called a seed in dicot plants is actually a caryopsis in monocot plants. Despite this distinction, most people refer to monocot caryopses as seeds, which will be done in this text. After the fertilized egg matures into a seed, the palea is the husk-like structure that surrounds the flat part of the seed adjacent to the rachilla, the stem that attaches the seed to the rest of the plant.

The outer husk-like structure that covers the rounded part of the seed is the lemma. Some lemmas have needle-like extensions from their tips that increase the seed's adherence to animals similarly to the burdock barbs mentioned earlier. These extensions are termed "awns." They may be short to long and are sometimes useful in distinguishing one grass species from another. Collectively, the pistils and/or stamens, lemma, and palea are referred to as the floret. Two husk-like glumes surround single or clusters of florets, providing additional protection while the flower and seed are developing within them. In other grass plants, like corn, the outer protective structures are simply referred to as husks. The style and stigma of corn are often referred to as silks, which is what they feel like when receptive for pollination.

As has been mentioned, flowers are sexual structures that produce seeds. Seeds usually develop from fertilized eggs, although there are exceptions where unfertilized eggs or non-egg cells form seeds. The process where seeds develop from non-fertilized eggs or non-egg cells is apomixis. Apomixis is not very common but has one advantage: it is not dependent upon pollination. Pollination can sometimes be sporadic. It is not a 100% dependable process. If a plant is dependent upon the wind for pollination

Figure 7.8. The tassel at the top of a corn plant is the male flower. The part that develops into an ear of corn is the female flower.

and there is no wind when it is ready to be pollinated, it will never be fertilized and won't produce seeds. Apomictic plants can produce seeds without fertilization; their ovaries turn into seeds automatically. The problem with apomixis is that it does not allow for genetic recombination. All offspring are genetically identical to their parents and contain all of the good genes of the parent and all of the bad genes as well. Additionally, this process slows down selection and evolutionary adaptation. Apomixis should not be confused with self-fertilization. A flower that can self-fertilize can use its own pollen and ovules to make seeds but they will still be sexually produced and have some level (albeit low over multiple generations) of sexually induced genetic diversity.

Flowers may be attached singly to a stem known as the rachis or in groups on the rachis. Groups of flowers on a single rachis are collectively called an inflorescence. The arrangement and number of flowers on an inflorescence are determined both genetically and environmentally. The three basic inflorescence structures are spike, raceme, and panicle. Flowers on spikes are attached directly to the rachis. Each flower on a raceme has a branch (rachilla) that attaches to the rachis. Panicles have multiple branches with flowers. Specific patterns of stem attachment may have additional names.

Some floral structures that appear to be single flowers are actually compound flowers. Species in the Asteraceae family (sunflowers and asters) often have two kinds of flowers on a single stem. The smaller flowers in the middle are usually yellow and may be mistaken for pistils or stamens, when they are actually numerous flowers, each with their own pistils and stamens. These inner flowers are known as disk flowers. The outer flowers, which may be mistaken as only petals, are actually sterile flowers known as ray flowers. They are the most colorful part of the compound flower to attract pollinators.

c. two plants in one

When you look at a plant, you are actually looking at one of two very specific stages in the life cycle of that plant. The tomato plant in your garden or the oak tree in the woods is called the sporophyte stage of the plant's life cycle. In angiosperms, the sporophyte contains every single cell of the plant except for the pollen and the ovules and the cells that create them. The sporophyte is considered an individual. The gametophyte is the second life cycle of the plant. The gametophytes are the cells that give rise to pollen grains and the ovules. These small groups of cells are considered separate individuals, produced by the sporophyte. The gametophytes will produce gametes and when these combinations go through sexual recombination, new sporophytes will be produced. The gametophyte phase is haploid, or has half the number of chromosomes (n) as the diploid sporophyte phase (2n). The "n" represents a complete set of unique chromosomes for each species. The two phases are collectively known as the alternation of generations. For some algae and other lower plants, these two phases result in separate plants, either gametophyte or sporophyte that live separately. For more evolutionarily advanced plants like the angiosperms, the two phases are physically connected. However, they are still considered separate organisms.

Figure 7.9. Sunflowers have two types of flowers, ray flowers on the outer edge of the flower attached to the petals and ray flowers throughout the floral center.

Plants such as mosses and liverworts have a pronounced gametophyte phase, with a less conspicuous, but visible sporophyte phase. The sporophyte relies on the gametophyte for some of its nourishment. Along the complexity continuum, ferns are most noticeable in the sporophyte stage, while the gametophyte is much smaller. Both stages are able to photosynthesize their own food, but most people would never recognize the fern gametophyte as a fern!

At the most advanced evolutionary level of angiosperms, only the sporophyte state of angiosperms is visible without magnification. At the gametophyte state, the few cells that create the pollen and ovules (known as the gametophyte) are found within and completely dependent for survival on the sporophyte. For all these plants, the mature sporophyte produces spores via meiosis and the resulting spores are haploid. They will develop into gametophytes. Meiosis is the process by which most living things create haploid cells, which are usually sexual cells. In plants, these haploids cells are not just cells but also organisms in their own right. In humans, the only haploid cells are the sperm and ova and these cannot survive independently of the person

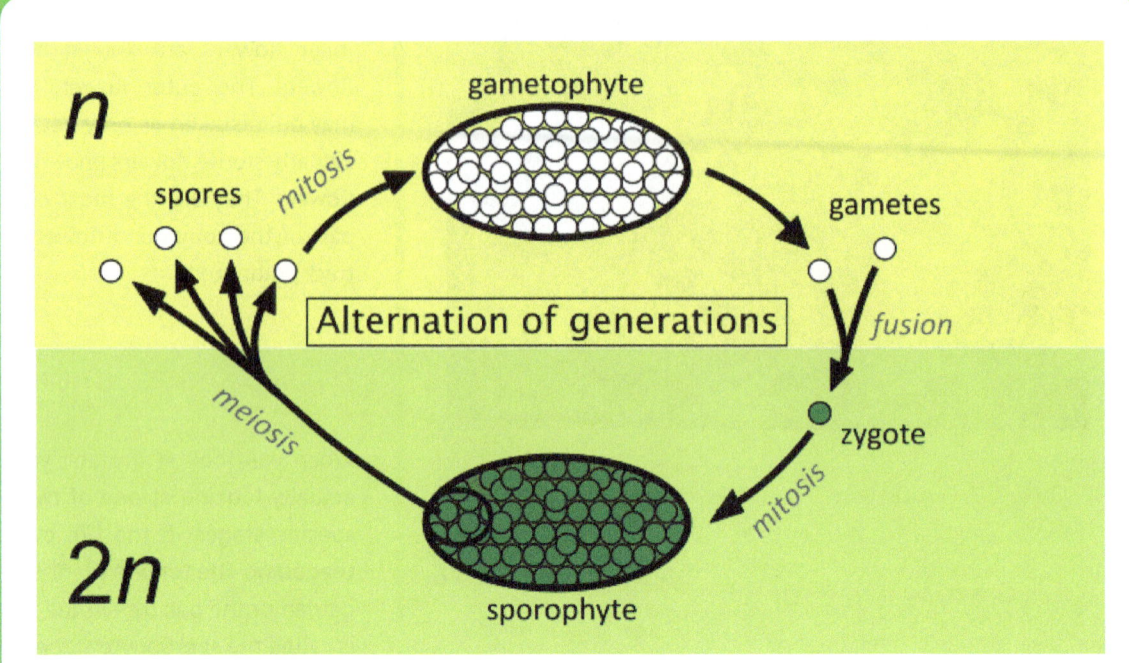

Figure 7.10. Plants alternate generations throughout their life cycle, moving from the sporophyte and gametophyte generations regularly.

who produced them. In fact, their lifespan may be quite short if fertilization does not occur. Meiosis is called a reduction division. In the process of meiosis, a single sexually dividing cell will start out with the 2n number of chromosomes; those will be doubled and finally divided up into four separate cells, each new daughter cell containing half the number of chromosomes (a haploid with "n" chromosomes).

The mature gametophyte produces gametes by mitosis. Since the gametophyte is haploid, the gametes will also be haploid. The process of mitosis is complicated but in short, a cell duplicates all of its DNA (the chromosomes that contain genes), splits the two copies of complete DNA apart and places them on separate sides of the cell, and finally splits the cell in two so that what was once a single cell is now two identical daughter cells. This process happens every time a new cell is produced, creating millions of carbon copies. This process occurs in you every day, anytime new somatic cells (body cells) are produced in your body.

When two single-celled haploid gametes from the same or different plants within the same species unite, they become a single-celled zygote. Since each gamete contributes one set of chromosomes (and is haploid), the zygote becomes diploid. The zygote undergoes mitosis to develop into a diploid sporophyte, where the life cycle begins anew with the production of spores.

d. people moving plants

Like other animals, people may move seeds incidental to their own movement. People may also intentionally move species from native or adapted habitats into areas where the plants become adapted to the new ecosystem without significantly affecting the existing life, where the plants are not adapted without human intervention, or where the plants overrun the existing ecosystem. Many plants grown in the United States today are not native to this country. This includes plants grown for aesthetic or nutritive purposes. Through breeding, humans have also altered wild plants into domesticated forms that would not likely thrive or even survive in an unprotected habitat.

Unfortunately, moving plants from one location to another may cause unintended and often negative consequerhonces. An increasing concern about invasive, non-native plants that may overtake native vegetation has led to the development of federal and state plant lists citing those species as restricted or noxious weeds. These lists vary among the states. Eradication efforts include complete plant removal, introduction of animals that will keep the invasive plants under control, and herbicides.

Kudzu (*Pueraria montana*) was introduced to the United States in 1876 at the Philadelphia Centennial Exposition. The government promoted kudzu for erosion control. Many farmers in the southern states planted it until the early 1950's when kudzu was removed from the list of acceptable ground covers. Kudzu was labeled a weed in 1970 and elevated to the Federal Obnoxious Weed List in 1997. Kudzu has spread aggressively throughout the United States and can kill other plants as it smothers and outcompetes them. In some areas of the country, kudzu is a major menace and covers thousands of acres. In addition to killing other plants, it can also take down power lines and damage buildings. Kudzu is found in over 7 million acres in the southern United States and can grow as fast as a foot per day. While some cultural and herbicidal controls have been effective, complete eradication has not occurred, resulting in this species dominating those locations in which it occurs, to the exclusion of much of the other vegetation. In addition to the obvious negative effects of kudzu, kudzu has been implicated in global warming. While it does not increase CO_2 levels, it does increase NO (nitric oxide) levels dramatically. Although only a secondary greenhouse gas, its plant-induced increase is of concern.

The Russian olive (*Elaeagnus angustifolia*) was promoted for use as an ornamental tree or a windbreak when introduced to the United States in the late 1800's. Although first grown in the central and western states, this species can be found throughout

Figure 7.11. Kudzu is an extremely aggressive invasive plant, growing quickly and smothering everything it comes across.

much of the country today. Russian olive grows more aggressively than some native vegetation to the point where the natural ecosystem is significantly disrupted. This affects the diversity of animal life as well as plant life. Some states now list it as a noxious or invasive weed. And unfortunately, it produces bright red berries that are a favorite of many birds. This allows for the rapid unchecked spread of this disruptive species.

A third example of a plant introduction gone awry is purple loosestrife (*Lythrum salicaria*). Although some plant material arrived as a contaminant in ships' ballasts in the early 1800's, other plants were intentionally propagated for their medicinal quality in treating digestive disorders and wounds or as ornamentals. The loosestrife used as an ornamental was originally believed to be sterile and thus would not be cross-compatible with the invasive loosestrife. Researchers at the University of Minnesota demonstrated in the 1980's that the ornamental cultivars would cross-pollinate with the invasive form of the species. Initially found along the Eastern United States, this species has inundated numerous marshy acres and waterways across the country. Purple loosestrife is listed as a restricted, invasive, or noxious weed on several state lists. Effective controls have been lacking but some success has been achieved with biological control (a beetle that when released, selectively consumes purple loosestrife and nothing else).

More than any other commodity, ornamental plants, grown for their aesthetics, have been moved across borders, mountains, and oceans. In North America, hundreds of non-native plant species have been introduced for landscape purposes from Asia, Africa, South America, and Europe. In most cases, these plants never become problematic when introduced in the new environment and they provide interesting and unusual landscaping elements. However, even non-invasive non-natives can spread. Most people have seen Japanese maple (*Acer palmatum*) in yards and lawns. Coming in both green- and red-leaved cultivars, the tree remains relatively compact and grows well in many different climates. However, if left to its own devices, it will produce viable seeds that will germinate and slowly spread. Because many of the most common Japanese varieties are red-leaved, it is easy to identify them when they pop up along wooded areas, roadsides, and nearby lawns. However, as they are slow to spread and easily removed, and their small seedlings may not survive, they are not usually considered an invasive species. Some of them have escaped into natural settings and reverted to green leaves, which makes them more difficult to spot.

Despite some of the negative consequences of moving plants from one part of the world to another, there have been many positive plant introductions. Corn is an excellent example. As a native New World crop, originating in Central to South America, corn has spread across the globe. It is now grown in dozens of countries, increasing their agricultural output and providing a relatively easily grown and cheap source of nutrition. Although many will debate the nutritive value of products such as high fructose corn syrup, corn does provide a stable source of nutrition for tens of millions of people. Apples, originating in the Old World, have also spread across the globe, providing a stable and important food source in many different countries. Other New World crops, such as tomatoes, potatoes, peppers, peanuts, avocados, blueberries, cranberries, cucurbits, and many others, have been cultivated successfully from Europe to Asia and help provide a highly diverse selection of fresh produce and processed foods.

While introducing plants into different cultures can have both good and bad impacts, one of the most mixed of results occurred with the movement of the potato. Potatoes originated in South America and were moved into Europe by the first European explorers. The potato was well suited to the Irish climate and quickly became a staple food for poor Irish farmers. It was easy to grow, store, process, and produced huge amounts of food cheaply. As a result, the Irish became dependent upon the potato, with estimates of up to one third of the population eating almost exclusively potatoes. Unfortunately, beginning in the 1840's, the disease late blight (caused by *Phytophthora infestans*) devastated the Irish potato crops. The almost complete failure of the crop led to the starvation deaths of millions of Irish poor. Without harvestable crops, many Irish left the country. Estimates suggest as many as two million people emigrated from Ireland, many of them coming to Canada and the United States. When the potato was introduced into Ireland, it was considered a boon, allowing the production of large amounts of cheaply produced food. However, when the crop failed year after year under the assault of late blight, the cultural changes were felt internationally as the Irish immigrated to America in overwhelming numbers, permanently shaping the American culture and society.

8

a. meristems

plants come in a wide variety of different shapes and forms. When we study the shapes of plants, we are studying plant morphology. In addition to the outward appearance of plants, we can also study the internal structures that make them function: plant anatomy. Plants are certainly not as complicated as animals, but they do have highly specialized structures that have evolved over millions of years, which have allowed them to spread and thrive.

One of the most important plant structures is the meristem. Meristems are unique to plants. The meristem is the part of the plant responsible for making new cells that ultimately develop into new plant parts. The simplest recognizable plant would have only two meristems. Both of these meristems would be called apical meristems, and one would be located in the very tip of the plant's root. The other would be located in the very tip of the plant's shoot. Newly germinated seeds have just these two meristems. However, plants quickly grow new meristems as they develop. The tip of every root and the tip of every shoot actively grows and enlarges because of the apical meristem present at that tip. A mature tree would have thousands of apical meristems between the roots and the shoots. The process by which the meristems makes new cells is called mitosis, described in the previous chapter.

The cells produced by the meristem are undifferentiated. This means that the cells don't have a particular fate—they don't know what they will become when they grow up. As these cells get older, the meristem continues to grow, and the cells behind the meristem start to elongate. They exist in the zone of elongation. After the cells elongate, they end up in the zone of differentiation. In this zone, they become different types of cells that perform specific functions. Although all the cells

Figure 8.1. The terminal bud on this Rhododendron is highly visible as the largest green bud. Three smaller axillary buds can be seen: two smaller purple buds next to the green terminal bud and the very tiny purplish bud directly underneath them sitting in the leaf axil.

the parts that make up the plant

59

Figure 8.2. Plants and plant cells are extremely flexible. An entire plant can be grown from a single cell. The process of generating new plants can be done in the laboratory using tissue culture in sterile environments and typically begins with a excised meristems.

started out identical, they will respond to different chemical signals, produce different chemical signals and as different genes are turned on in individual cells they will turn into different plant parts.

The apical meristem in a root is relatively simple. The meristem grows behind a root cap. The root cap is a pile of cells that the meristem wears like a hard hat. The cells in the cap are constantly lost and break off as the root pushes through the soil. The root cap protects the meristem and is constantly replenished by cells coming up through the root and produced by the meristem. In the opposite direction, the root meristem is producing new cells to elongate the root.

The apical meristem in the shoot is a bit more complex than root meristems. The shoot apical meristem is contained within a bud when it is dormant. The meristem is surrounded by very young leaves (called leaf primordia). Then the leaf primordia and meristem are secondarily contained under hard leaf/bud scales that protect the bud from freezing during winter. When the buds emerge in the spring, the leaf primordia break through the hard bud scales, expand rapidly, and the branch starts to elongate as the meristem produces new cells.

In addition to apical meristems at the top of every branch or twig, axillary buds exist in every location where a leaf is attached to a branch. These are called axillary because the junction between a leaf's petiole (the short stem that holds the leaf to the branch) and the stem is called an axil. These buds may never come out of dormancy. Their primary function is to act as a reserve or a backup source of leaves and new branches if the terminal apical meristem/bud (terminal means the bud at the end of the branch) gets broken or damaged. In addition to terminal apical buds and axillary buds, some plants will produce buds directly from the branches if all of the leaves are cut off. Rhododendrons are a common landscape plant that undertake this type of growth. Rhododendrons are very easy to grow because they do not require many inputs, and if they are badly damaged, they recover quickly. If you cut every single branch off a rhododendron, it will often put out a whole new set of branches before the summer is over. Although this is not recommended, they can be very aggressively pruned on a regular basis.

Other meristems also exist. When a leaf primordium enlarges and develops into a new leaf, meristems called leaf marginal meristems allow for this growth and expansion. However, these meristems shut down when the leaf reaches its maximum size, and the leaf will no longer grow any larger or repair damage that may occur after it has reached that size.

Grass plants are very different from most other plants in form and have different types of meristems. When a grass plant is mown, the leaf blades recover and grow back. This is because grass plants have a meristem about halfway down the leaf, called an intercalary meristem. If the top of the leaf is removed but the intercalary meristem stays in place, the leaf will continue to grow. If the intercalary meristem is removed, the grass leaf will stop growing, and the plant will put out new leaves to replace the damaged leaf. In addition, grass plants don't have typical apical shoot meristems (although their root meristems are just like those in other plants). The primary shoot meristem in a grass plant is called a basal meristem (although some people will call it an apical meristem). It is located low, in the crown of the plant at the soil level, and all new leaves originate from this point. Unlike other plants that have apical meristems at the tips of their branches and produce new leaves continually, this meristem has a limited lifespan and can only produce a small number of leaves, usually less than ten. The grass flower also arises from this meristem. Once the grass plant flowers, it will not produce any new leaves and will be replaced by new meristems from the same location if it is perennial and die if it is annual. These new meristems, or buds, are effectively new plants called tillers.

When a tree dies and falls over in the woods, it may remain on the forest floor for years. The reason it takes so long to decompose is that tree trunks are made of lignin. Leaves and green tissue are mostly composed of cellulose that, unlike lignin, is very easy for microorganisms to degrade.

The meristems we have discussed so far are all primary meristems—that is, they are the first meristems to be produced and they are all "green." Roots are not generally green, but we can usually separate primary and secondary meristems by their color. Primary meristems produce green or white tissue and do not produce appreciable levels of wood. A secondary meristem does just that: It produces lots of wood. Wood is a structurally complex material composed of random molecular arrangements of lignin. The fact that wood is composed of a hard material like lignin and that the molecules are randomly arranged at the molecular level makes wood very hard and resistant to decay. Lignin gives plants strength and allows for rigidity. Primary meristems do not make much lignin but do make cellulose. Cellulose is also rigid but is a much simpler molecule with much less strength.

Secondary meristems are also located within the plant. Only plants that produce wood have secondary meristems. A plant that does not produce wood is called herbaceous versus woody and will typically stay green throughout its lifespan. It may also be an annual (it only lives for one season). Secondary meristems are also considered lateral meristems because they make a plant wider in a horizontal direction, while the primary meristems grow vertically. The two secondary meristems are the vascular cambium and the cork cambium. A tree will start out with just primary meristems, but as it gets woody, the vascular cambium will start to produce wood (secondary xylem and phloem), which will allow the tree to get wider and wider and

61

increase the water-moving ability of the tree, as well as allowing it to get taller. The cork cambium allows the tree to produce new bark as the volume of the tree increases. The word cambium means meristem but is only used to describe the special secondary meristems.

b. it all starts in the roots

Roots are the foundation of all plants. Without a root system, few plants can survive for very long. Roots serve three distinct purposes: 1. to anchor plants in place; 2. to absorb water and mineral nutrients; and 3. to store carbohydrates and plant reserves in a safe location—free from herbivores and most pests.

When a seed germinates, it puts down new roots in the location that it germinates. Unless is it transplanted by humans, the plant will spend its entire existence in that one spot. If the soil is deep, well drained and suited for plant growth, the seed will thrive. If the soil is bare and rocky, the plant will likely struggle (although some plants are well adapted to rocky locations and poor soils). In either case, the roots will hold the plant to that place and keep it from blowing away in storms and washing away in flooding rains. The roots will provide support for the heavy trunk and canopy and allow large trees to stay upright as they annually increase in height and bulk.

Tubers, bulbs and corms may all look the same but their structure is very different and the tissue they arise from are different. A tuber (like a potato) is actually derived from stem tissue. Some tubers are derived from root tissue. And even though a bulb grows in the ground, it is a compact stem, as is a corm. The arrangement of parts in a corm and a bulb, however, is very different ad that is what separates one from another.

If the seed does germinate in a poor area for growth, it will need the contribution of every possible root in order to absorb enough water and mineral nutrients to survive. Although plants can absorb a marginal amount of water through leaves, the bulk of the water they use for all of their biological functions comes from the roots. For a short time, plant roots can be removed, and stems can sometimes be placed in water to keep the plant alive, like cut flowers. However, it is ultimately the roots that are best suited to absorbing and moving water into the plant. Without a root system, plants die. Even plants that can survive long periods of time with their stems in a container of water will develop roots and transition to them for water absorption. In addition to water, almost all of a plant's mineral nutrients also are absorbed through the roots.

Figure 8.3. Dandelions are ubiquitous in the landscape and manage to survive so well because of their deep and difficult to remove taproots.

Figure 8.4. Most of the mushrooms people find in the woods have evolved to degrade lignin. Woodchips in urban landscapes can provide and excellent moist environment for them to access lignin.

Finally, roots store the bulk of any perennial plant's food reserves. Cortex cells do not undertake photosynthesis and have the available volume to store large amounts of starches. In addition, because roots are underground, it is difficult for herbivores to access them and steal reserves from the plants. Some plants even go to extra lengths to hide resources underground in the form of tubers, bulbs, and corms. These structures persist from year to year, allowing the plant to hide its food supply with little risk of discovery.

Not all plants have roots. Spanish moss (*Tillandsia usneoides*) is a plant commonly encountered in the Southern United States that absorbs all of its water and nutrients from the air or from the surfaces of the leaves of other plants as it hangs on to other plants. And it is not actually a moss at all; it is a flowering plant. Spanish moss is a type of epiphyte, an organism that lives on plant leaves but does not consume those leaves. It just lives on the surface.

Different types of roots are better suited to different environments. Roots can usually be described as fibrous roots, tap roots, or a combination of the two. Fibrous roots tend to move only shallowly into the soil, but spread out over large areas. Tap roots can grow very deeply, but are localized to one small area. Each type of root has advantages and disadvantages, depending upon the environment in which it is growing. Many weeds, like dandelions (*Taraxacum officinale*), have tap roots, making them difficult to kill and allowing them to be perennial. Many grass plants have fibrous roots that allow them to capture water and nutrients at the top of the soil and hold together like a dense mat. For this very reason, turfgrasses can be harvested as sod. If turfgrasses had tap roots, they could not be harvested in large carpets.

There are even many different types of roots that do other things besides serving the typical three roles of roots. Mangrove (*Rhizophora* spp.) plants produce extremely specialized roots, called prop roots, which hold the plants upright and anchor them in marshes and tidal areas. Without these specialized roots, plants would wash away readily. However, these roots actually aid in the formation of new land masses, as they slowly trap silt and soil. Mangroves are also extremely important for slowing the effects of coastal erosion and protecting inland areas from tidal surges. When mangrove swamps are destroyed, inland areas become much more prone to flooding during storms. Corn (*Zea mays*) also produces prop roots. Corn is a very tall and slender plant and requires additional roots that develop above the soil level to hold the plant upright. Contractile roots are another type of specialized root, often produced by corms. These roots actually serve to pull the corm deeper into the ground so that it cannot be eaten by passing herbivores and is less likely to be damaged by frosts.

Roots generally all have a similar structure. As has been mentioned before, the very end of the root contains the root cap that protects the root apical meristem. Behind the meristem are the zones of elongation and differentiation. Further along the root is the root proper, the functional and typical part of the root. The entire root is surrounded by an epidermis, or skin, made up of a layer of epidermal cells that allows most things to pass in and out of it freely. In the zone of elongation, all the cells within this epidermis are a blank slate—they don't have a functional purpose yet. In the zone of differentiation, the different types of cells are defined and start to function.

The bulk of the functional root is made of cortex cells. Cortex cells run from the epidermis to the stele. These cells store carbohydrates and starches and allow the movement of water and mineral nutrients deeper into the root, usually in the spaces between the cells (called apoplastic spaces). These cells are loosely packed and rather generic. In the center of the root is the stele. The stele is the central core of the root and is much like a long tube contained within the root. The stele is covered in a sheath of cells called the endodermis (the stele is not a distinct structure but the collection of structures in this part of the root. This layer acts like a second inner skin and controls the movement of water and nutrients in and out of the stele much more actively than the epidermis. Underneath this layer is the pericycle. Although not a meristem, the pericycle is where new roots are formed and branch from old roots. The last structures in the stele are the xylem and phloem. These two structures act like a highway, transporting water (xylem) from the roots all the way to every leaf and transporting sugars (phloem) from the leaves to every root.

The gymnospersm are also known as the conifers: Remember, gymnosperms arose earlier in evolution than the flowering plants, which are either a dicot or a monocot.

The last major root feature is the root hair. Roots produce root hairs to dramatically increase the amount of water and nutrients they can absorb. Root hairs are extremely fine hairs that are produced from root epidermal cells. These hairs are typically only found at the very end of a root, near the zone of differentiation. Root hairs tend to have a short lifespan and do not cover the entire root surface.

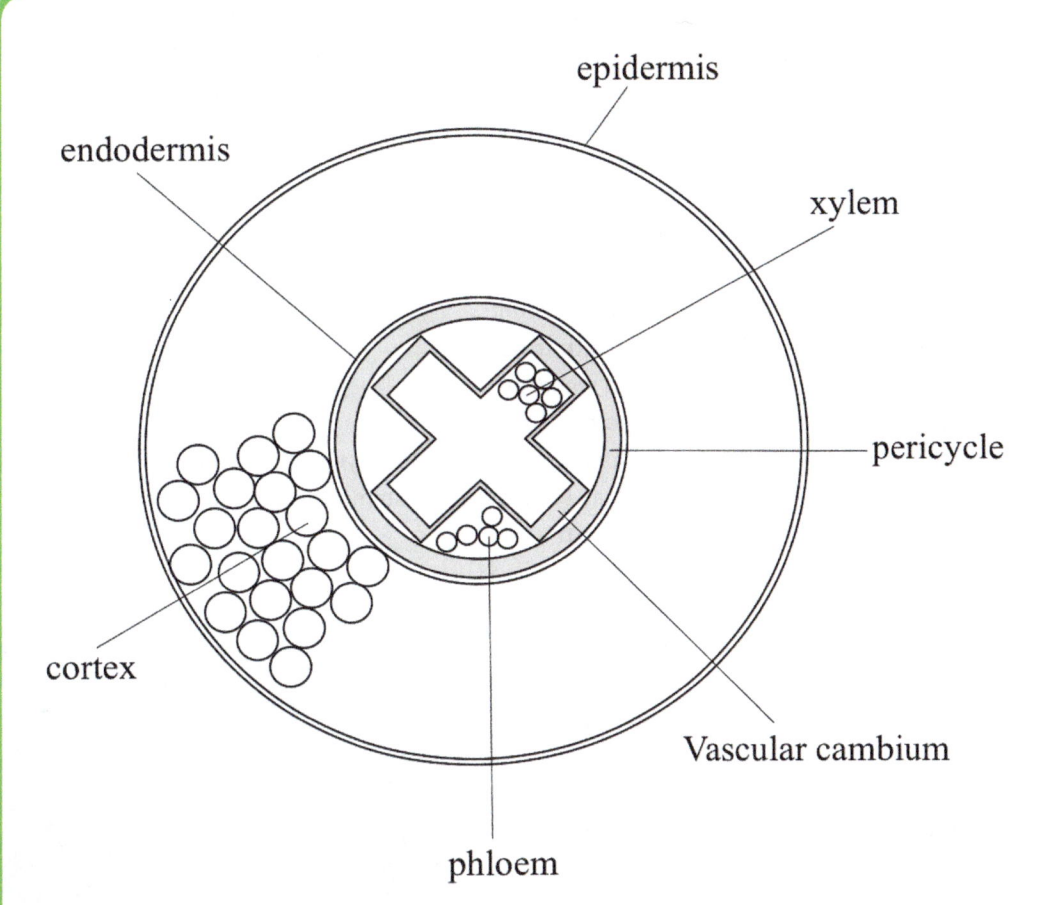

Figure 8.5. The roots of dicots are highly structured and contain seven primary tissue types (note that in the diagram, all of the tissue types are composed of individual cells that operate in unison but cells are only depicted for the cortex, xylem and phloem). Monocot roots are much less organized and the xylem and phloem are distributed throughout the cortex.

c. always growing upward

The roots are very important to any plant, but the roots need the other plant parts to be successful. While roots usually handle the water and mineral nutrients, roots do not have chloroplasts and cannot produce their own food. They are completely dependent upon the leaves to produce sugars and the stem to move the sugars to them, using vascular bundles of xylem and phloem.

While the roots of most plants are generally very similar, there are significant differences between types of stems. Differences in stems exist between monocots and dicots and between herbaceous or woody plants. As has been mentioned previously, herbaceous stems are mostly green and do not produce any wood. These stems only have primary meristems and are limited in how tall or wide they can become. Herbaceous stems have an epidermis—just like roots—but the stem epidermis contains chlorophyll and is therefore green, unlike the root epidermis. Also similar to roots, cortex cells exist underneath the epidermis. So do xylem and phloem vessels, but instead of one central stele, multiple bundles of xylem and phloem circle the outer edge of the stem in dicots. In monocots, these vascular bundles are randomly distributed throughout the cortex. In the middle of a dicot herbaceous stem is an area known as the pith, similar in composition to the cortex. Because there is not a distinctly different central area in monocots, pith is not a structure they contain. Not all plants develop wood, but those that do are always dicots or gymnosperms. Some monocots may appear woody, but they just contain very hard-packed layers of primary tissues—not true wood.

People who plant apple trees are often disappointed by how long it takes the trees to mature and produce apples. While some dwarf types might produce fruit after 2 years, 4-5 years is more likely depending on growing conditions. Standard, full size, apple trees may take 10 years to produce their first fruit!

A woody plant produces wood because it has secondary meristems, the vascular cambium and the cork cambium. The outermost layer of a woody plant is the bark, which is also called the cork. Some woody plants produce very thick layers of cork, while other produce very thin layers. The cork used in wine bottles comes from the cork, or bark, of the cork oak (*Quercus suber*). The cork cambium constantly produces new cork to replace old cork. Cork is the primary protective structure of a woody plant. The cork layer prevents excessive amounts of water from leaving the plant (using a waterproof material called suberin- notice the species epithet of cork oak!) and keeps insects and pathogens out. It also acts to insulate the plant to some degree. As you travel into a woody plant, the next layer is the secondary phloem. This is produced by the vascular cambium. The next layer is the vascular cambium, followed by the secondary xylem that goes all the way to the center of the woody plant. When a woody plant first germinates and establishes, it only has primary meristems and is not woody. As it matures, it develops secondary meristems and becomes woody. Most of what we call wood is actually old secondary xylem. Trees and woody shrubs produce new wood in the spring and summer and produce annual rings that can be seen when a tree is cut. Because rings are only produced during the growing season (springwood in the spring and summerwood in the summer), the age of the tree or shrub can be determined by counting the rings. As the plant ages, the oldest xylem dies and becomes a repository for plant waste products. This

Figure 8.7. Only dicots and gymnosperms produce true woody tissue. Although a palm tree appears to have bark and wood, it's unorganized monocot tissue does not produce rings like dicot trees and its wood is just highly compressed parenchyma tissue.

is called the heartwood (as opposed to the sapwood, which is still an active part of the vascular system). Additional channels, called rays, run from the most newly produced xylem and go inward to the older dead xylem. The vast bulk of any tree is actually dead xylem, which provides the necessary support to hold the tree up as it climbs into the sky to reach sunlight.

Wood is an evolutionary advantage that some plants have come to rely on. While the ability to produce wood allows a plant to compete well against other plants (by raising its leaves high into the sky and maximizing light absorption), wood also requires a lot of energy to build. When a plant relies on wood as a part of its lifestyle, it needs to initially shift resources away from other important functions, such as reproduction and leaf production. Plants that produce wood typically produce fewer seeds earlier in their lives. However, once they have established themselves, they will persist perennially, and if they reach a suitable size, they can often outcompete annual plants that may emerge below them.

Not all stems follow the typical pattern of growth. Two very important types of stems don't look like stems at all. Rhizomes (below ground) and stolons (above ground) are stems that grow parallel to the ground (below and above) and spread out away from where the plant originated. These structures allow the plant to establish itself in new locations without using seeds. Turfgrasses frequently use both, which explains why lawns can spread and thrive even when they start out in only a small area. Strawberries *(Fragaria × ananassa)* are another very common plant that produces stolons, allowing a few plants to turn into a large patch. When the rhizome gets long enough or the stolon finally touches the ground, a new plant develops in that location and eventually produces its own rhizomes and stolons.

d. the antenna for the sun

As we've already read, chlorophyll is an antenna for the sun. As the analogy goes, it is appropriate, but we could also call the leaf the antenna for the sun. It is, after all, the structure that contains the chlorophyll, and leaves will often move their position in relation to the sun to maximize the amount of sunlight they can collect, much like a radio antenna can be moved around to get a better signal.

Because the main purpose of the leaf is to collect sunlight, most leaves have evolved with that function in mind. For the most part, leaves are light and thin. Because they are light, they do not weigh down the plant and are hard to remove. In a windstorm, most leaves will remain attached even at winds reaching hundreds of miles per hour. When water hits a leaf, it will not collect on the leaf surface, nor will it not knock the leaf off the plant. The connection between leaf and stem is strong yet flexible and normally only breaks if the plant is getting ready to drop its leaves anyway, as it begins the process of winter dormancy.

Leaves are generally also very thin so that as much as possible can get to all of the chlorophyll and chloroplasts contained within the leaf. Leaf structure is very similar to that of a sandwich. The top and bottom of the leaf are composed of a single layer of translucent cells called the epidermis. These cells are covered in a waxy substance called cuticle that keeps water from evaporating out of the leaf and keeps diseases and other invaders out. Because waxes are hydrophobic (they are resistant to water), any water that lands on the leaf will bead and quickly run off. Inside of the leaf is a layer called the mesophyll. The top part of the mesophyll is called the palisade parenchyma. Parenchyma is a word that means a generic type of cell. Cortex cells in the roots and stem are also parenchyma—it just means the majority type of filler cell. Palisade parenchyma is important, however, because it contains the bulk of the chloroplasts and absorbs the majority of the light that comes into the leaf. These cells are stacked tightly and are oriented upright to maximize light absorption. The

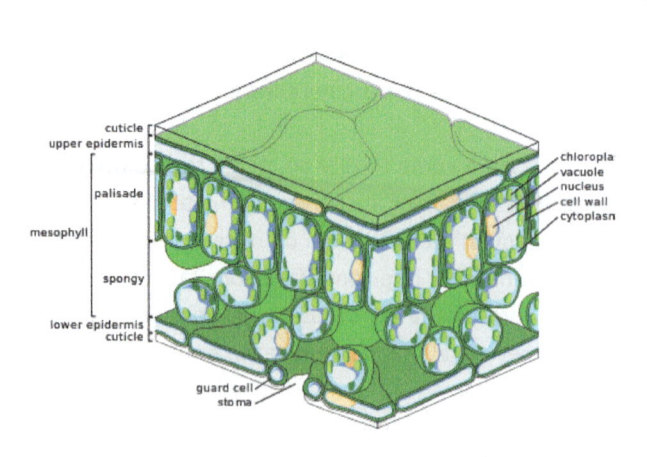

Figure 8.6. The interior of a leaf resembles a sandwich, with epidermal layers containing and protecting the inside structures.

layer of cells below the palisade layer are the spongy parenchyma. These cells absorb some light, but the primary role of this layer is gas exchange. Any carbon dioxide that comes into the plant comes in through this layer and any water vapor that needs to leave will exit through this layer. It is called spongy because these cells are very loosely packed. Inside the spongy parenchyma is also the xylem and phloem vessels.

In order for gases to move in and out of the leaf, they need an entry or exit hole. The lower epidermis of leaves is covered in holes called stomates. Stomates open and close with the help of guard cells. Two guard cells surround every stomata, and when the plant has a lot of water, the guard cells open as a result of the excess water pressure. When the plant is low on water, the guard cells automatically close. The ability of the guard cells to automatically open or close helps the plant conserve water during drought but they continue to function even when the plant is fully hydrated- that is, they don't only work based on hydration cues. Stomates will also close during the night when no photosynthesis is going on and plants do not need to exchange gases.

Anyone who has seen a leaf will notice that it is covered in veins. In some plants, these veins are fanlike and spread across the leaf surface. In other plants, they are parallel, such as in irises and grasses, and follow the contour of the leaf. These veins contain the xylem and phloem that run throughout the plant. The xylem and phloem are embedded in the mesophyl and long distance transport of water and carbohydrates throughout the plant only happens in the veins. However, cells that are not directly connected to the veins will move water and sugars short distances through or around other cells to get them to the nearest veins, which act like highways for transporting materials over long distances.

Not all plants have leaves and not all leaves look like leaves. As plants have evolved in different climates and different environmental conditions,

Figure 8.8. Spines are how cacti protect themselves from herbivores in the desert. The spines are actually modified, nonphotosynthetic leaves.

Figure 8.9. While cacti modify leaves to form spines, roses modify epidermal stem cells to form prickles.

they have made significant changes to their structures that have allowed them to succeed. Cactus plants are an excellent example of the evolutionary modification of leaves. Cacti do not have typical "leaves." Instead, they have sharp spines. These spines, however, used to be leaves millions of years ago. As cacti adapted to arid environments, spines gave them protection against predators seeking to eat them and steal all of the water they had accumulated. Cacti use the bulk of their body, the stem, to photosynthesize and use their modified leaves (spines) to protect themselves. Holly (*Ilex* spp.) plants are another good example of a similar modification. Although holly leaves do look like leaves, the end of every leaf has a spine, a sharp point that protects the plant. This adaptation is just a partial modification, but it may explain how cactus plants evolved slowly over millions of years.

Some plants produce other structures similar to spines that are not derived from leaves at all. Thorns are actually a modification of the stem, appearing at the ends of branches. Prickles are a modification of the stem growing from epidermal cells on the stem. Although people talk of roses (*Rosa* spp.) having thorns, they technically have prickles.

Other types of leaf modification serve a very different purpose, specifically for climbing. Plants that grow low to the ground will often try to climb on other plants or upright structures so that they can gain access to more sunlight. Tendrils are produced by many different plants but are common among cucurbits (cucumbers, pumpkins, and squash) and are a modified leaf that allows the plant to grab on to stakes, walls, or other plants and climb.

Although not strictly a modification, stems and roots can often grow from the wrong place on a plant. These types of tissues are called adventitious. When a stem starts growing a root, it is an adventitious root. Tomatoes (*Solanum lycopersicum*) will commonly produce lots of adventitious roots, especially if you lay them on their side and let the stem touch the soil. Ivy (*Hedera* spp.) plants will climb many different surfaces to get closer to the sun, but instead of

Figure 8.10. Ivy can climb great distances up trees and buildings using adventitious roots that cling to surfaces as the plant climbs.

using tendrils, they produce adventitious roots that grow along the length of the stem and cling to surfaces. These roots can be so strongly embedded in walls that they can rip building facades off when the ivy is removed. While plant morphology and anatomy generally follow predictable patterns, there are hundreds of different modifications that exist that allow plants to thrive and succeed in unique lifestyles and challenging environments.

9

a. plants are chemical factories too

in the last chapter we learned about all the structures of plants, how they work, and how they allow plants to thrive. These structures, such as leaves, roots, and stems, are critical to the success of any plant, but learning plant structure is just one part of understanding plant biology. The science of living structures is called anatomy or morphology—in essence, the shape of living things. Morphology is easily observed with the naked eye or low-powered microscopes. Morphology can be complicated but it is generally much less complicated than all of the intricate biochemical processes that go in inside plants. The science of examining the chemical reactions that all plants need to function is called physiology. Just like animals, plants undergo thousands, if not millions, of chemical reactions every second. These chemical reactions are responsible for everything a plant does. While the results of these reactions can sometimes be observed with the naked eye, often the results are difficult to tease out from the myriad other reactions surrounding them. The science of plant physiology is the science of how plants work.

Although it was not mentioned in the context of plant physiology, an extremely important chemical reaction has already been discussed in detail: photosynthesis. And while photosynthesis occurs in thykaloids, contained in grana, inside chloroplasts, dispersed through cells that ultimately comprise leaves, the process of photosynthesis is a chemical reaction and the subject of plant physiology. And while photosynthesis is the most import plant chemical reaction, a plant could not survive if all it did was photosynthesize. Other important chemical reactions include building new molecules for new structures, transcribing and translating DNA into proteins, signaling mechanisms, making defensive chemicals, and sensing and responding to their environment, just to name a few.

Plants are chemical factories, just like people, and the wide range of molecules they produce and their interactions with each other is astounding. But plant physiology is not just chemistry. It is a complicated science and it crosses into many other fields including molecular biology, cell biology, genetics, physics and sometimes even plant ecology!

b. signals and plant hormones

People interact with their world primarily through their senses: sight, hearing, touch, smell, and taste. These senses are controlled by muscles and nerves and sensations are transmitted through neurons chemically and electrically to the brain, where those signals are then converted into other signals that our brain can recognize and respond to. Plants do not have brains or neurons. Despite this limitation they still interact

making the plant work

with their environment on a daily (if not hourly basis). There are obviously many differences between how animals and plants respond to stimuli. In addition to the lack of a nervous system, plants often respond to a stimulus in a localized way. Signals can move throughout an entire plant, but this movement is often slow, requiring many hours or days. Most of the signal response that does occur in plants happens at the cellular or tissue level and individual cells and tissues respond accordingly.

The most widespread signaling mechanism used by plants is hormones (phytohormones) or plant growth regulators. When hormones are produced or move to different parts of the plant, they cause a plant response. They usually generate a response by affecting which types of genes are turned on or off and at what levels. When genes are turned on, proteins are produced that can have profound effects on plant growth. The hormones themselves respond to many different signals but the two most import signals are light and gravity. When a plant is

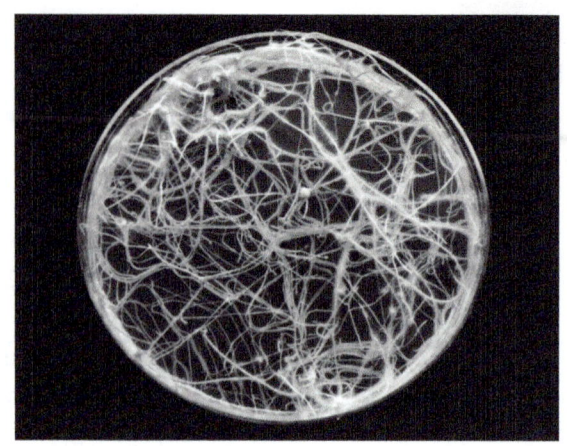

Figure 9.1. Different plant hormones can have major impacts on plant growth. In this case, an elevated level of auxin is placed in a growth media and germinated tomato seeds will fill a dish with roots but will not produce any other tissue type.

placed in the dark for a long period of time, it becomes yellow and thin, or etiolated. Plants need light and when they are deprived of light, they grow quickly without fully developing so that they can reach the light and start photosynthesizing. When they are starved for light, the hormone gibberellic acid (GA) is produced. When cells come into contact with GA, they become very long. The result is yellow plants climbing faster toward the sky to find more light. If an etiolated plant is exposed to light, it will slow its growth and turn green as chlorophyll is produced and the plant starts producing carbohydrates. Unfortunately, once a plant has undergone etiolated growth, it will likely never recover fully, especially if it is an annual plant. It will always be weak and spindly and prone to falling over. Gibberellic acid does more than just cause cell elongation, however. It also plays a key role in seed germination by triggering the conversion of starch stored in the seed into sugars that the new seedling can use quickly and it exerts an influence over many of the other developmental plant pathways.

Gibberellic acid was one of the first of the plant hormones to be identified. Unusually, it was not discovered in a plant! Rice (*Oryza sativa*) is a major crop in many Asian countries. The high temperatures and humidity in many of these regions are excellent for rice production, but the climate can also stimulate many diseases. Rice plants are susceptible to a disease called foolish seedling disease. After the plants germinate, they grow very quickly and get thin, tall, and yellow. Unfortunately, these plants cannot hold themselves upright and often collapse and die without ever producing any harvestable seeds. Although Japanese scientists had discovered the fungus that caused the disease in the 1890's, it was not until the 1930's that they discovered that the reason the pathogen caused such odd symptoms was that it produced a plant hormone. The pathogen had learned to manipulate the plants it infected by producing a chemical naturally found in the plant at low levels. The fungal pathogen that causes the disease is *Gibberella fujikuroi*, so they called the plant hormone gibberellin.

Another critical plant hormone is auxin. Multiple variations of auxin exist but the most common and most import auxin is indole-3-acetic acid (IAA). While gibberellin primarily effects cell elongation, auxin primarily affects cell division. That is, when auxin is present, more cell division will take place. However, not all cell types respond to all plant hormones and different cell types (root vs. shoot) will respond to hormones in different ways. Auxin is a major factor in new root development. If high auxin levels are present, more roots will be produced. This fact is often used in horticultural settings. When plant cuttings are used to produce new plants, they start out as just cut branches with no roots. In some species, treating a plant stem with auxin will cause the formation of new roots. Once roots have been produced, the cutting is now a new plant and can be transplanted into a

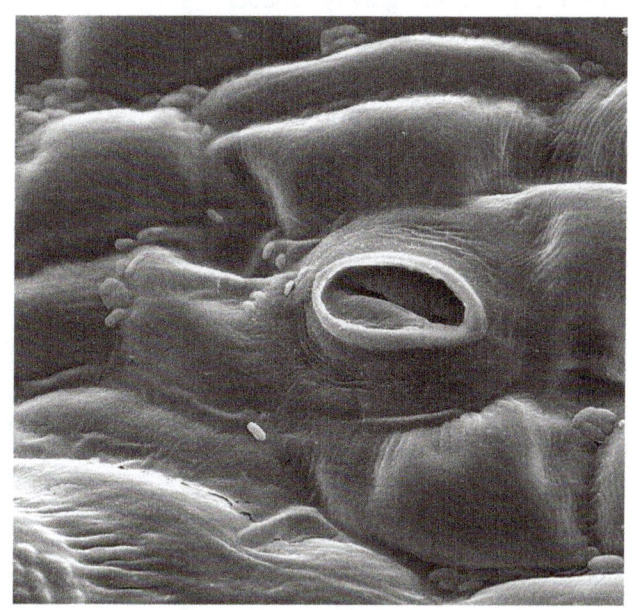

Figure 9.2. Stomates present in plant leaves allow for the exchange of gasses between the plant and the atmosphere and open and close to increase or decrease gas exchange.

greenhouse or a field (horticulturalists usually use synthetic auxins produced in the laboratory that often produce a stronger response than natural auxin). Conversely, auxin prevents the formation of new shoots. All plants have many meristems (or axillary buds) along each of their shoots or branches. However, only a few of those meristems actually develop into new branches or leaves. In the phenomenon of apical dominance, existing leaves produce auxin and send it downward toward the root system. As long as auxin is being produced and flowing downward, few if any new leaves and branches will develop along that particular shoot. If the topmost leaves are cut from the shoot, auxin flow will halt and new meristems will begin developing. Once they have matured they will produce their own auxin and stop even more meristems from developing. Horticulturalists will frequently use this fact to change the shape and development of plants. By pruning perennial plants regularly, the plants will become "bushy" as they produce new branches and become denser. When apple trees are pruned, the removal of the apical meristems will cause new branches to be produced that can allow for more apples and even make the tree stronger than if it had few very long branches.

The other major role (and there are many minor roles for auxin in the plant) that auxin plays is in the response to light, gravity, water, and touch. Plant growth in response to light is called phototropism. Plant growth in response to gravity is called geotropism or gravitropism. Plant growth in response to water is called hydrotropism. Plant growth in response to touch is called thigmotropism. "Tropism" means "to grow." In many cases, plants are responding to all of these environmental cues simultaneously, but plants will often respond more strongly to one cue than another.

Roots are positively geotrophic: they grow toward gravity. Shoots are negatively geotrophic: they grow away from gravity. Auxin is responsible for telling the plant that roots grow down and shoots grow up. If a plant is placed parallel to the ground, auxin will collect along the side closest to ground. Shoot cells in this accumulation zone will start to divide and elongate, causing the shoot to turn away from gravity. Root cells in this accumulation zone will not grow, the cells that are on the other side of the root will divide and elongate, and the root will bend downward. Most people have seen plants growing toward the sun. It seems obvious that since plants require sunlight, they will do their best to orient themselves in the direction of the sun. The phenomenon of phototropism is

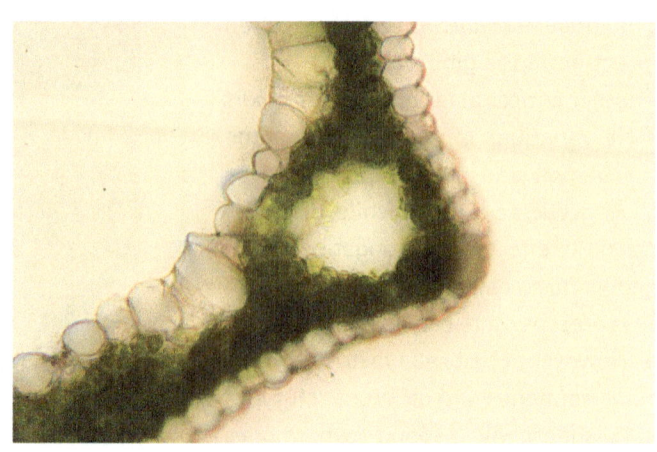

Figure 9.3. All leaves contain xylem and phloem for water and carbohydrate movement. The orientation of the veins depends on the type of plant. In grasses, the veins run parallel to the leaf, as can be seen in this cross-section of a grass leaf blade.

similar to geotropism. Shoots are positively phototrophic, growing toward the sun, and roots are negatively phototropic, growing away from the sun. When sunlight hits a shoot, the plant responds by transporting auxin to the side of the shoot facing away from the sun. When the auxin accumulates in high enough concentration, cells in the side facing away from the sun start dividing and elongating, thus bending the plant toward the sun. The phenomenon of hydrotropism is much less understood than these other phenomena. While it has been demonstrated in the laboratory that plant roots can move toward water and this movement is mediated by auxin, geotropism usually has a stronger influence on plant roots, making hydrotropism difficult to observe in the soil. Thigmotropism happens when a plant responds to touch and is most commonly observed when tendrils and vines attach and climb along other structures. When a cell from a specialized climbing structure contacts a solid surface, it produces auxin. This auxin is transported to the other side of the structure (like a tendril) and the accumulation of auxin results in growth that is observed when the tendril grasps and climbs up the surface. Shoots do not generally exhibit a thigmotrophic response unless the shoot is a vine. Roots, however, can be negatively thigmotrophic. That is, they will often grow away from contact. This allows roots to grow through the soil, avoiding rocks and other impediments.

Cytokinins are other common plant hormones. Cytokinins interact frequently with auxin and can have many complicated effects, but their primary role is in the development of new shoots and buds through cell growth and differentiation. While auxins are produced in the leaves and move downward in the plant, cytokinins are produced in the roots and move upward in the plant. Consequently, while auxins stimulate root growth, cytokinins stimulate lateral shoot growth. Because cytokinins and auxin interact with each other, sometimes doing opposite things, the concentration of one versus the other is critical in the regulation of plant growth and development. Finally, cytokinins can help slow the process of senescence (aging) and allow plant tissues to remain active and growing for longer periods of time. And horticulturalists can also use cytokinins for agricultural purposes. Applying them to crops in the field will sometimes result in larger crop yields. This is also true for certain crops and gibberellin, particularly certain grape varieties when GA application results in more, larger grapes.

The next major plant hormone is ethylene. Ethylene is a little unusual because it is not a solid molecule in the plant, it is a gas. In fact, it is a hydrocarbon and burns quite well. Ethylene has three major functions in plants. Firstly, it is responsible for cell thickening. Plants exposed to high concentrations of ethylene will have thicker stems and have a better ability to stand upright. The most common example of this effect is when plants experience regular moderate to high winds. These plants respond to windy conditions by producing more ethylene that results in thicker stems. Plants grown in greenhouses often have weaker, thinner stems because they do not experience windy conditions and thus do not respond to that environmental cue. Secondly, ethylene speeds fruit ripening. When fruits are shipped over large distances, they must be picked before they are ripe, otherwise they will spoil before they ever make it to market. However, unripened fruit may not be ready for market when it arrives at its destination. In an effort to get properly ripened fruit into the hands of consumers, it will often be treated with ethylene in a warehouse or shipping truck just before sale. This process ensures that fruit will be ripe for sale. Surprisingly, as fruit ripens it produces its own ethylene. If a rotted or overripe piece of fruit is placed in a container with unripe fruit, the unripe fruit will ripen more quickly. Not all fruit is susceptible to ripening by ethylene, however. Finally, ethylene is responsible for the loss of leaves in the fall, as perennial plants become dormant. As auxin levels decrease in the fall with colder weather, ethylene production is increased and leaves break from the branches at a thin layer called the abscission zone. Ethylene also speeds the process of senescence, opposite to the effects of cytokinins.

The final hormone among the major five is abscissic acid (ABA). It was originally thought that abscissic acid was responsible for degrading the abscission layer in petioles that then allowed the leaves to drop in the fall, hence the name. It turns out that this is not always the case, as ethylene plays the primary role in leaf drop. Abscissic acid is produced in leaves and buds when plants go into dormancy, but its primary role is to slow the growth of plant structures and help prepare them for winter metabolism. It is also responsible for triggering the changes that close stomates in the leaves when plants are under drought conditions. Finally, abscissic acid is responsible for establishing the dormancy of newly produced seeds so that they do not germinate prematurely (such as during the winter) and die.

Additional plant hormones do exist (at least half a dozen more) but they tend to have fewer effects, their effects do not appear to be critical to plant function, or they are simply not well understood. Some of these plant hormones have extremely specific effects, such as salicylic acid (the same thing as in aspirin) and jasmonic acid, which are both responsible for turning on a wide array of plant defense mechanisms to defend against disease.

Finally, it should be noted that plant hormones are typically active at extremely low concentrations and their ratios are critical for plant function. This is most easily demonstrated when plants are grown experimentally, particularly plant cells and tissue in petri dishes (also known as sterile plant tissue culture). While a small amount of plant hormone will bring about the desired effect, too much plant hormone may completely halt that same effect or in some cases, kill the plant cells outright.

c. moving water

Plants require water to undertake just about all of their basic functions. The majority of plant water is absorbed through the roots and moved to the rest of the plant. But the process of moving water is not always simple and requires multiple steps along the way. Water moved into the roots is passive. That means that energy is not generally required to move water into plant roots, it just happens naturally. However, in order for water to move into the plant on its own, a delicate balance between the soil and the roots does need to be maintained by the plant. Water moves from one location to another by a process called osmosis. Osmosis is a property of water; it is a rule of the physics of solvents (things that other things are dissolved in) and solutes (the stuff dissolved in the solvent), not necessarily a biological operation but critical to biological function. In the process of osmosis, water moves from an area of low solute (or salt) concentration to an area of high solute (or salt concentration). When water does this it is trying to make the concentration of the solutes (or salts) equal between the place it is coming from and the place it is moving to. This works really well in roots. As long as the plant keeps the roots filled up with more salts/carbohydrates/solutes than the soil that the root is growing in, water will just naturally flow into the root from the soil. However, if the roots cannot maintain their solute concentration or if someone dumps a ton of salt into the soil (such as fertilizer), the roots will stop absorbing water and start leaking it into the soil, resulting in severe wilt.

When water moves into the roots, it generally does not move into the actual cortex cells but moves between the cells. This area is called the "apoplasm" or the "apoplastic spaces." The water will migrate all the way to the center of the root and cross the endodermis, which was discussed in the last chapter as an impenetrable wall that protects the xylem and the phloem. Water will be actively transported across the endodermis from cells next to the endodermis and into the xylem. Water can then move into the rest of the plant. This is called root pumping and happens all day and even during the night.

Water moves all across the plant from the xylem but this is another passive process; the plant does not put energy into it but the sun does. The primary mechanism for water movement throughout the plant is called transpiration or evapotranspiration. Transpiration relies on evaporation from the leaves. As the sun heats the leaves, water in the leaves evaporates and exits through the stomates. But once that water has left, it leaves a vacuum—the absence of water. Water molecules are strongly attracted to other water molecules because they

Figure 9.4. The color of fall foliage is the result of pigments present in leaves. They are present all year long and help absorb light energy but only become present in the fall when plants stop producing green chlorophyll.

are polar (have two sides) and thus each water molecule acts like a little magnet. When water molecules are attracted to each other, hydrogen bonds are formed. Water has strong cohesion; that is, water attaches to water. Plants use this to move water far up the stems. When water evaporates out of a leaf, water from nearby takes its place. Because all the water molecules are attached like a chain, it pulls all the water next to it. When every leaf is transpiring, water will be pulled up the xylem from the roots, even if the distance is hundreds of feet.

When the sun goes down, plants shut their stomates and transpiration stops. Plants can control transpiration by opening and closing stomates during the day also—if too much water is being lost, stomates will close. When plants cannot keep enough water from leaving and the soil is too dry to take up more water, plants wilt. If they wilt too badly, eventually they will not be able to recover and will collapse and die. Occasionally at night, particularly with small plants, water can still be pushed up the plant by root pumping and droplets will appear out of the hydathodes of leaves. These are called guttation droplets. Hydathodes are small pores on the edge of leaves. Once the sun rises and transpiration restarts, guttation droplets will be sucked back into the plant. On cloudy or humid days, water will transpire out of leaves very slowly. Because nutrients are carried to every part of the plant in moving water, this will slow the rate at which plants get their nutrients. Some plants may even start to turn yellow if water movement stalls for an extended time, because they are not receiving continuous nitrogen to produce more chlorophyll and carbohydrates.

d. responding to temperature and day length

The two primary cues that plants respond to are temperature and light. Temperature is important to plants because they do not have a mechanism to generate heat; they are not homeostatic like people—that is, they cannot maintain a constant temperature. Plants rely on the external temperature to influence their biology. When it is cold, plants will slow down and not produce new tissue. If it is too cold, plants may freeze and experience significant damage. When it is hot, plants will grow faster and produce more tissues. If it is too hot, plants may literally cook to death. Because plants can't maintain their internal temperature to a constant optimum, they need to be able to cope with temperature extremes.

We've just mentioned transpiration in relation to water movement and it turns out that transpiration also plays a role in plant cooling. When water is converted from a liquid to a vapor, heat is required. This is seen when a pot of water boils: heat is converting the water from liquid to vapor. When people sweat, heat from their bodies converts the moisture into vapor. And when people sweat, the heat leaves the body with the vapor and cooling results. The same thing happens with transpiration. As water in the leaves is converted to vapor, heat exits with vapor, cooling plants down. When humidity shuts down transpiration, not only can plants suffer from nutrient deficiency but they can often overheat and go into heat stress. Transpiration can easily drop the surface temperature of leaves by five to ten degrees Fahrenheit. And as more water is pulled from the soil, the cooler water in the soil will continue to cool the rest of the plant as it moves through the stems and branches, replacing warmer water.

In the winter, plants can be susceptible to freezing. Although annual crops will gradually decline with cooler weather and often die with the first or second hard frost, perennial plants need strategies to deal with freezing temperatures. Plants contain significant amounts of water, and when water freezes inside plant tissue, it can cause permanent and sometimes lethal levels of damage. When water freezes inside cells it can cause major membrane damage. Ice formation will cause cytoplasm to swell and ice crystals can literally pop cells. In less-severe cases, water can freeze in between the cells. This usually will not cause popped cell membranes but it can still cause tissues to break apart and cause water loss out of cells. Plants cannot always entirely prevent cell freezing, but they do a very good job of it through the process of hardening. When a plant hardens itself for the winter, it actually changes the composition of its cells. When a plant is actively growing it needs a lot of water. But water becomes a liability during the winter because of its propensity to freeze when it gets cold. Plants prevent freezing in a manner similar to that of a car's cooling system. In a car, liquid coolant prevents the engine from overheating. But if a car uses water as its coolant, it will freeze in the winter. As a result, cars use a mixture of water and other chemicals that have a lower freezing temperature. This ensures that a car's coolant will remain liquid well below freezing. Plants cannot go to the local auto store and buy coolant, but they can remove water from their tissues. As cooler temperatures abound, plants start replacing water in their cells with carbohydrates. Carbohydrates act as a solute and drop the freezing

Figure 9.5. Poison hemlock closely resembles the common carrot in the appearance of its leaves and flowers. The poison produced by the hemlock (not to be confused with the tree called hemlock) is a secondary metabolite alkaloid. Poison hemlock is most famous for its use in executing the Greek philosopher Socrates.

temperature of the water. One of the reasons seawater does not readily freeze is that it is full of salt. The salt is a solute that lowers the freezing point of water. Whereas drinkable water freezes at 32°F, seawater freezes at 28°F. The carbohydrates that plants produce are even more effective at reducing freezing temperatures. A well-hardened plant can easily survive at 20°F without freezing and some plants can survive without freezing at temperatures of 10°F or below.

Cold temperature is not always a bad thing for plants. While annual plants typically respond to cold temperatures by dying, many perennial plants need and expect cold temperatures to undertake or complete their annual life cycle. The name for this phenomenon is vernalization. Plants that require vernalization require a steady and prolonged period of cold before they will break dormancy and produce flowers and then seeds. Plants that utilize this strategy ensure that winter has indeed passed and that it is safe to start growing again. It also ensures that the plants will not flower in the fall, only in the spring after a cold period. If a plant flowered in the late summer or fall, it might produce seeds that would germinate too late in the season to survive the winter. By flowering in the spring, the seeds that are produced will have the summer to grow and develop.

Just as waking up from dormancy is dependent upon temperature and light, so is going into dormancy. When temperatures drop, perennial plants go into dormancy. Dormancy is a physiological stage where the plants are essentially asleep. Metabolism is very slow and no new growth occurs. Dormant plants remain dormant until temperatures rise and/or light levels increase. Most people believe that temperature is the only factor that wakes plants out of dormancy. In fact, light is often also necessary. Some plants do respond almost solely to temperature; forsythia, for example. But using only temperature as a method to wake from dormancy can be a problem because temperatures can vary widely throughout the season. Very warm days can occur even in the middle of winter. If a plant wakes up and starts producing leaves in the middle of the winter just because of a few warm days, it is likely to freeze and suffer significant damage once the temperatures drop. Because the seasons are controlled by the Earth's position in relation to the sun, the amount of light present in the winter is very low. As spring approaches, light levels increase. While temperatures fluctuate daily, light levels always increase or decrease at a slow and consistent level. Plants that measure light levels *and* temperature can protect themselves much better from coming out of dormancy early than plants that respond only to temperature.

It is this mechanism of measuring light that also triggers leaf drop in most perennial plants. The specific molecules that detect light changes are called phytochromes. These molecules are protein pigments that absorb light. Once light has been absorbed, their structure changes and this change is a signal to other plant systems that depend on and respond to environmental light cues, through the production of hormones (leaf drop, dormancy, flowering, etc.). The response to changes in light is called "photoperiod."

While some plants require vernalization to flower, some plants depend solely on day length for flowering. Long-day plants flower when the amount of light they receive exceeds their particular light threshold, usually in the spring or early summer. Short-day plants flower when the amount of light is below their particular threshold, usually in the summer or late fall. Day-neutral plants do not depend on light and may require vernalization or simply flower when they get old enough to flower. Plants are extremely diverse and the ways in which different species respond to environmental cues is also extremely diverse.

e. secondary products

Plants produce a large number of highly varied molecules and products. The topic of plant biochemistry falls into the realm of plant physiology, but it is simply too large a field to cover very well in this text. Some plant biochemistry has already been discussed, specifically photosynthesis, hormones, etc. But plant chemicals are even more diverse than that.

Plants produce two very broad categories of molecules: 1. Those required for basic metabolism and growth and 2. Those that serve secondary, nonessential purposes. Those molecules required for basic metabolism include everything the plant needs to survive: structural molecules, enzymes, hormones, carbohydrates, polymerases, etc. If a plant cannot, for some reason, produce a molecule in this category, it will die or it will be weak and easily outcompeted by other plants in a natural environment. These molecules and products are essential. Secondary products are not essential. If a plant does not produce a particular secondary product, it may have no effect on the plant at all.

So why do plants produce secondary products at all? Even though they are not essential, many of them can still help increase a plant's survivability. Not all of their purposes are known, however, and some secondary products don't appear to have any useful purpose for the plant at all. Secondary products are often related to defense. Although they can help the plant survive an attack by an insect or a disease, they are not required for plant function. Other secondary products have medicinal value for humans but do not appear to directly help the plant. Many of these chemicals are alkaloids. Alkaloids like caffeine, nicotine, atropine, and morphine have a wide array of human uses and can be medicinal at low doses (but often toxic at high doses). Alkaloids like solanine serve no useful purpose for humans; they are just poisonous. It is possible that many of the poisonous alkaloids have evolved over time by plants to prevent consumption by herbivores (people, deer, sheep, et al.). Other secondary products include things like tannin. Tannin or tannic acid produced by plants protects them from insect feeding and can combat fungal infection. Humans use tannins to flavor food (such as wines) and to preserve, or tan, leather. Many secondary products exist and different plants will produce a wide diversity of these compounds. Many are still to be discovered.

10

a. weather is what it's all about

<u>not all plants</u> grow best in every environment. In fact, every plant has a particularly environment that it is best suited for and in which it will grow to its full potential. Over the course of the evolution, those that survived changes in specific environments became the most common and most successful plants in those environments. And those plants that survived were the ones that produced the most offspring, as we have already mentioned. The Earth is constantly changing. Although we do not always see changes that take place over millions of years, these changes influence plant populations and evolution.

Weather can be perceived in different ways. Locally it may be raining and cool today and dry and hot tomorrow. But weather can also occur regionally, in specific climates. Weather within a particular climate usually has a repeatable pattern or is predictable over an extended period of time. One way to look at climate is to divide the earth into temperature zones. The general categories of climate zone include the frozen zone, where no plants grow; the tundra where a limited number of species are adapted to short growing seasons and long, cold winters; the temperate zone, where four distinct seasons (winter, spring, summer, and fall) occur each year; the subtropical zone, where the four seasons are not distinctly separated by dramatic climate changes (but where chilling temperatures may occur); and the tropical zone surrounding the equator, where temperatures and day length remain fairly constant throughout the year. Within these zones, however, are often smaller regions that may be different from the rest of the zone. High mountainous areas that occur in a temperate zone will be more similar to a frozen or tundra zone and the appropriate types of plants will survive there.

Figure 10.1. Plants that are adapted to arid climates develop strategies to cope with the desert environment. The most recognizable of these plants, the cacti, utilize their stems for water storage during long periods of drought.

how to grow plants

Much of the contiguous 48 United States of America would be classified as temperate. Alaska has areas that are frozen, as well as some that would be considered as being in the far-north temperate zone. The southern states have subtropical zones, while Hawaii would be classified as a tropical zone. If you travel from one zone to another, driving east to west (zones difference also follow elevation changes which can be east to west, as the example of mountainous areas provides) or north to south, you will notice that the plants change. Plants that grow in Texas may not grow in Maine. If they do, their growth pattern will be very different. The climate has a dramatic impact on which plants grow where and how they grow.

Figure 10.2. Growers are completely at the mercy of the weather for the success of their crops. Any number of climactic factors can destroy an entire year's harvest.

Arid regions tend to have plants that survive the long dry periods by collecting and storing water throughout much of the plant during the rare rainfall events to support their growth during the dry periods. Some plants in arid regions germinate quickly and may live only during the few days or weeks in which water is present.

Humid areas receive ample rainfall, but plants that are not adapted to substantial amounts of water often suffer more disease problems than plants that normally grow in these wet regions. Tropical regions often receive more than 40 inches of precipitation annually and some areas receive hundreds of inches of rain per year. The native plant life in tropical regions is dominated by dense forest (called jungles in tropical zones). Much of this forested land has been cleared for human habitation, including farming, commerce, and housing.

A large portion of the middle of the United States of America receives

Figure 10.3. Pivot irrigation is a relatively water-efficient method often used to grow crops in arid environments.

roughly 15 inches of rain per year, a relatively small amount of rain. This area is known as the prairie. Grass species were the dominant plants before humans settled on the land and began farming. Trees were uncommon, since there was not enough rain to support their growth. Early settlers made their homes using sod cut from the prairie where they lived.

Much prairie acreage has been converted to agricultural farmland, including parts known as the corn or wheat belts, where thousands of acres of these crops are harvested annually. Where the rainfall is most limiting, supplemental irrigation must be utilized. Pivot irrigation systems are especially noticeable when flying over these regions, because they produce green circles of vegetation during the growing season, while the surrounding area is various shades of light green to brown.

Climactic differences and the associated weather separate different geographic regions (Texas is generally hot and dry and Maine is generally cold and often wet) but as mentioned before, weather occurring on a daily basis can also influence plant growth. In fact, the short-term weather changes over periods we can see and experience, have the greatest significance on the chances for a plant species' survival in nature, as well as in a managed landscape or in agriculture. Temperature extremes and rainfall are two of the most critical aspects of weather that plants must adapt to in order to ensure a species' survival.

Of all the things that make a farmer or grower successful, the weather is the most important. If the spring happens to be too wet and cold, a farmer cannot plant seeds early. Because the seeds are planted late and the crop starts growing late, his crop may freeze in the fall before it can be harvested. If a late frost occurs in an apple orchard, it may kill all the apple flowers and the trees will not produce any fruit that year. The same is true for plants in nature. A wet spring may be bad for some plants while a dry spring may be bad for others. Growth patterns, reproduction, germination and establishment of any plant are all subject to the whims of the weather. Most people may not realize it, but farmers insure their crops just like any other person would insure their car or their house. But a farmer is usually insuring their crop against the weather. From hail to tornadoes and floods, many different weather events can damage or destroy a crop. And when a crop is destroyed, it cannot be easily recovered. A crop is not like the panels of a car that can be replaced or have dents repaired. Once a crop is lost to the weather, the farmer loses all of the income that croup would have generated. As a result, wise farmers will insure their crop if possible so that if it is destroyed, they will receive reimbursement for that crop so that they will not go bankrupt or lose their farm.

Most crops are grown outside. Because the weather can be highly unpredictable from week to week or month to month, the weather is the farmers greatest impediment to success. If the weather is good for the crop, the farmer will succeed. If the weather is poor, the farmer will fail.

b. not all soil is the same

The word 'soil' has different meanings, depending on one's viewpoint of the usefulness of soil. When it gets stuck to the bottom of our shoes, we usually call it 'dirt'. When it supports plant life, it is considered soil. Soil is the place where most plants grow. Some plants are grown in water (hydroponics) and some plants are grown in gravel or in artificial materials but most plants need soil. The majority of the nutrients a plant receives from its roots are

In addition to just being a sand, silt or clay, every soil has a name. Soil scientists have designed special names to delineate particular characteristics of different soils. One type of the many hundreds of soils is an Enfield silt loam. This is different from Bridgehampton silt loam which is different from a Paxton fine sandy loam!

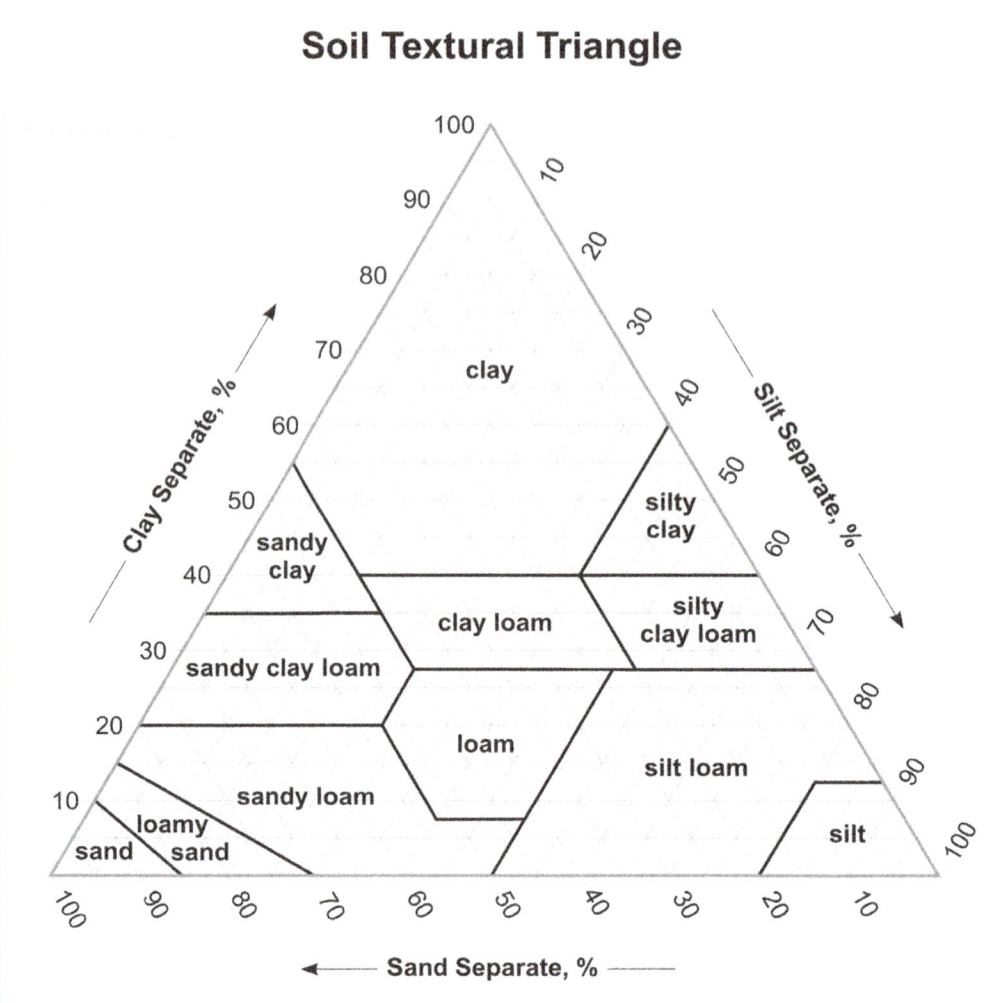

Figure 10.4. Soils fall into 3 main textural classes: sands, silts and clays. A loam is often considered the optimal soil for plant growth but not every loam is exactly the same.

stored in the soil, as is the majority of the water. The soil really is a lifeline to the plant and if a plant is grown in a poor soil it will never thrive.

Soil is composed of solid matter, as well as air and water contained in the pore spaces surrounding the solid matter. For the best plant growth, the ideal soil is usually 50 percent solid material and 50 percent pore space. Water and air each occupy half of the pore space, or 25 percent of the total volume of the soil. However, this ratio is not constant. As the soil warms up or cools down and as water evaporates from the soil or it rains, the proportion of air water will change. When there is too much water in the soil, plants can actually suffocate- they cannot take up enough oxygen to function. If there is not enough water in the soil the plants will wilt.

The solid material of soil is divided into the organic and the inorganic portions. The organic matter is "organic" because it contains carbon and because it has come from living things, from the decomposition of plants and animals. Fully decomposed organic matter is known as humus. Organic matter is very important to plant growth. Organic matter in the soil holds on to water and nutrients very well and stimulates additional decomposition and a healthy soil community. Communities of soil organisms are responsible for nutrient cycling and soil stability. These soil organisms include bacteria, fungi, nematodes, insects and a whole collection of invertebrate animals. All of the these organisms live and die in the soil and help keep it functioning.

The inorganic part of soil contains minerals from the erosion of rocks over many thousands and millions of years. The size of the inorganic material determines its classification. The smallest particle size, less than 0.002 mm in diameter, is clay. The largest particles are classified as sand, while silt is the material larger than clay but smaller than sand. Sand is further subdivided, also based on particle size. The sand categories include extra fine sand, fine sand, sand, coarse sand, and extra-coarse sand.

The inorganic components of soil determine the texture classification of a soil. The percentage of sand, silt, and clay varies based on the classification. That is, when you mix different amounts of sand, silt, and clay together, you get a different type of soil. A diagram, known as the soil texture triangle, shows the various names given to soil textures. For optimal plant growth, loam is considered the best soil texture. The amount of sand in loam provides good soil drainage after a rain, while the amount of clay provides nutrient and water-holding capacity. Some soils are awful for plant growth, particularly pure clays and sands. A pure clay will constantly be waterlogged and suffocate plant roots. It will also compact very badly. Clays can be used to line a pond, they work well at holding water in. A pure sand will constantly be dry, and starve plants for moisture. In addition, sand will not hold onto nutrients well and plants growing in sandy soils will often be yellow and/or stunted.

Soils are so important that the US Government has a whole group of people working on the preservation and improvement of soil: the National Resources Conservation Service

The soil in which a plant grows is critical for its success. Plants receive all of their water and nutrients from the soil, and the soil provides support for the plant. Many plants will struggle if placed in the wrong soil. And the characteristics of soil go beyond just physical characteristics, as we'll see in a minute. Plants often evolve to prefer certain soil types. For this reason, forests or particular types of trees may be common in an area—those trees like that soil! However, if a tree is placed in a soil it is not adapted for, it will struggle and may eventually die. Farmers grow certain crops based on soil characteristics. Some crops do very well in soil that contains very high levels of organic matter, whereas many crops may not mind how much organic matter is present. Often, greenhouses use artificial soils that work really well in the artificial environment of the greenhouse but would do very poorly in nature.

c. to fertilize or not to fertilize, that is the question ...

A soil is more than just its parts. That is, just because two soils may be considered a "loam," or just because two soils have five percent organic matter, that does not mean they are equal. In addition to the sand/silt/clay composition of a soil or the percentage of organic matter, soils have varying levels of nutrients that may be stable or based on changing plant life and climate. The activity of microbes in the soil will have an impact on the quality and characteristics of a soil. The way a soil is used (forest, farm, field, fallow) will also impact a soil. Soils have an intrinsic pH that can be very stable or change with the seasons. And even the mineral or organic components can be variable. Sand is not all made from the same stuff. A soil that contains sand high in calcium is very different from a soil that contains sand high in silicon. Even organic matter comes from different places. Decomposing grass roots are very different from piles of sawdust turned into the soil. As you can see, soils are complicated, and soil scientists spend much time teasing soils apart and building better soils.

It should be apparent that good-quality soils are critical to plant growth. Unfortunately, if a soil is of poor quality, very little can be done to fix that soil in a short amount of time. Changes and improvements to soils (like increasing organic matter) take place over decades or centuries. And new soils cannot be created in human lifespans. So every farmer or grower does their best to maintain the quality of their soil and to make the soil better, if at all possible. Often, this may include practices that increase organic matter or microbial activity in the soil. Sometimes this is done by adding raw plant material to be broken down naturally over time. Sometimes this is done by reducing tillage to slow decomposition processes and slow soil erosion. But no matter what is done to a soil to make it better, all soil needs fertilizer if it is to be used for growing agricultural crops.

Figure 10.5. When the soil is low in nutrients, plants can demonstrate a wide variety of different deficiency symptoms. Yellowing of leaves (chlorosis) is a common symptom of low nitrogen.

The topic of carbon fixation and carbohydrate production has already been discussed, but this is only part of what plants need to grow and thrive. In addition to carbohydrates, plants need raw materials in the form of inorganic nutrients. The nutrients are in the form of soluble ions (charged atoms that can float around and dissolve in water, just like table salt) present in the soil. The ions are absorbed through the root systems and move upward in the plant to all the different parts of the plant. Usually, nitrogen is considered the most important plant nutrient, followed by phosphorous and potassium. These nutrients all do different things within the plant and are used by every cell in the plant. These three nutrients are called macronutrients because they are required in very large quantities. Additional macronutrients include sulfur, magnesium, and calcium. Sometimes carbon, hydrogen, and oxygen are also considered macronutrients, but these three nutrients are always present in the air or soil. Rarely does a plant have a deficit of oxygen unless it is submerged under water, in which case oxygen becomes a very important nutrient to a suffocating plant! For this reason, some people will call oxygen the most important nutrient. Silicon is also considered a macronutrient by some scientists.

If a plant does not have enough of the right nutrients, it will grow slowly and weakly or may not grow at all. Plants that suffer from nutrient deficiencies can be yellow or brown, have spots, be short and stunted, or have many other different symptoms that result from being starved. In natural settings, the nutrients and other soil characteristics present in the soil will influence which plants can grow in that soil. Plants have different preferences and needs and will thrive in some locations and fail in others. Many times weeds are found in areas with low levels of nutrients; the reason that weeds are so frequently successful is that they require few nutrients and can grow almost anywhere. In agricultural settings, growers utilize fertilizers to compensate for an improper nutrient balance or the lack of nutrients. In fact, modern agricultural crops rarely can be grown without fertilizer.

Most soils have very low levels of macronutrients, especially nitrogen, particularly if they have been used for agriculture in the recent past. If there is no natural process for nitrogen cycling, nitrogen is used up quickly by both plants and soil microbes and leaches through the soil quickly. In order to produce large plants with many leaves and large fruit (or grains, etc.), crops need to be regularly fertilized. Some plants are adapted to lower levels of fertilizer or have special tricks to make their own, but the vast majority of crops needs to be fertilized if they are to produce a yield. Many of the crops grown today were actually bred to be reliant on fertilizer. Early farmers often did not have the ability to fertilize their crops. This is one of the reasons why soils became "played out." After just a few years, soils can run out of the nutrients naturally present and will no longer sustain a crop unless nutrients are replaced. The use of fertilizer in modern agriculture and the ability to breed plants that can use all that fertilizer is the primary development that has allowed us to grow the huge amounts of food necessary to sustain the world's population.

The question is not really whether to fertilize or not, but how much fertilizer to use and what type. When it comes to the amount of fertilizer used, too much can be a bad thing. Fertilizers are really just salts. When you give a plant too much salt, it burns. If you pour salt into the soil, plants cannot take up water and will also burn or wilt to death. When plants are fertilized, growers give the plant just enough to get a good crop, but they are very careful not to overfertilize. Not only can they damage the plants with too much fertilizer, but fertilizer is expensive. Overfertilizing is a waste of money. And too much fertilizer can damage the environment. When excess nitrogen and phosphorous are applied to a farm, they often leach or run off the farm as it rains. Over time, they can migrate into lakes, rivers, and the ocean and disrupt these systems, allowing algae and other organisms to proliferate and consume all of the oxygen in the water. When this happens, many of the fish and other marine life can die.

Fertilizers also come in many forms. The most common form is synthetic fertilizer. This frequently comes as a powder or a granule that is made in a factory. Usually, synthetic fertilizers are made from natural gas, which is processed into a form that can be applied to the soil. Synthetic fertilizers are relatively pure and easy to use. Because the fertilizer is concentrated in this form, a lot of fertilizer can be stored in a small area. In addition, precisely controlled amounts of fertilizer can be applied. Unfortunately, many synthetic fertilizers rely on fossil fuels (because they are made from natural gas). If there were no fossil fuels, these fertilizers could not be produced very easily or in large quantities. And it is easy to make mistakes with synthetic fertilizers by overapplying them. Another common type of fertilizer is organic fertilizer. These fertilizers are organic because they contain carbon and come from some recently living source. Cow manure is a very common type of organic fertilizer. These types of fertilizers are often inexpensive, but they are very difficult to apply and transport, often smell, may take a long time to release nitrogen into the soil, and have a highly variable level of nutrients.

There are many types of organic fertilizers and while not all of them are as difficult to use as cow manure, they all have some limitations. Usually, it is just very difficult to get enough organic fertilizer at one time to cover very large areas of land, especially when crops are grown on thousands of acres. And organic fertilizers often have very little nitrogen, although they can slowly increase the organic content of soils when applied regularly.

Fertilizers can also be divided up into other categories, such as water soluble (they dissolve easily), water insoluble (they do not dissolve well), granular fertilizers, foliar fertilizers, etc. Each type of fertilizer has a specific use and may be more suited to one situation than another.

d. just one of many meanings for organic

During the Green Revolution, industrial farming took off. With the ability to fertilize and irrigate large amounts of land and the development of crops that could thrive in that environment, growers could produce unprecedented amounts of food. But industrial agriculture relies on chemicals, both in the form of fertilizers and in the form of pesticides. Most people do not realize where their food comes from or how it is grown, but often

Figure 10.6. Greenhouses are used to extend the season of many different crops. Some crops are grown almost exclusively in the greenhouse or in special high tunnels (which are similar to greenhouses), particularly in cool regions with short growing seasons.

people will become concerned when they find out that pesticides have been applied to their food. In the United States, the government regulates the maximum amount of pesticide residue that can remain on crops when they are consumed. In order for consumer crops to be as safe as possible, the levels that are allowed are very, very small. In addition, the government approves the types of chemicals that can be applied to foods. Over the past 30 years, the types of chemicals that can be used in crop production have been rapidly shrinking and newer chemicals are much safer than old chemicals, with much less human and animal toxicity and minimal potential to cause environmental damage.

Organically grown crops are an alternative to growing food with pesticides and synthetic chemicals. While we have discussed the term organic to mean "containing carbon," food that is grown organically is food grown without synthetic inputs. Organic crops can still be grown on huge farms in an industrial setting, but only organic fertilizers can be used and no synthetic pesticides can be applied to the crops (although there are a few exceptions). Organic crops do still require some level of input. Organic does not mean going back to the 1700's and growing crops "the old fashioned way". Many of the agricultural techniques of the pre-Green revolution can be very destructive to soils and the farm land. Organic growers utilize science and technology to grow crops but rely on methods of pest control and fertilization that are often considered "natural". Although the United States Department of Agriculture (USDA) does regulate the use of the term organic in food labeling, some food-related terms can be misleading. People often like to buy "free-range" eggs, assuming that the chickens that laid the eggs spent their lives frolicking in woods and fields. However, the USDA defines a free-range chicken as one that can spend sometime outside of it's coop. Thus a chicken that spends 15 minutes in an outdoor cage the size of washing machine is technically considered "free-range".

Organic crops do have a few drawbacks. Specifically, organically grown food tends to be more expensive than nonorganic food because it requires more labor to grow. For example, when crops are not grown organically, herbicides can be used to keep the weeds out of the field. In an organic crop, herbicides cannot be used, so some type of manual labor will be required to keeps the weeds down and labor is often more expensive than chemicals. An additional issue with organic crops is the quality of the crop. Synthetic pesticides exist for just about every pest, meaning that crops that can have pesticides applied will be aesthetically perfect. Crops grown without chemicals will often have blemishes, spots, and even insect holes. None of these imperfections should hurt the consumer, but people are used to buying perfect apples, and organic crops are not always aesthetically perfect. In addition, organic foods may go bad more quickly than traditionally produced food because they are not treated with preservatives. This can be problematic if you don't intend to consume the food in the next day or so. And finally, a concern that sometimes arises is whether any crop can ever truly be organic. If an organic crop is produced using cow manure but the cow was heavily injected with antibiotics, those antibiotics are now introduced to the soil and in the plants growing environment. Many soils where organic crops are now grown used to be conventionally farmed and the chemicals that were once used on those crops are still present in the soil. Although these concerns are not great, they do pose interesting questions about the organic nature of organic crops.

Some plants are easily transplanted while others are more difficult to move. Not only does the type of plant determine the ability to transplant it but also its age. A tomato plant can easily be moved anytime during its life but an oak tree cannot be reliably moved once it has established itself. If an older oak tree is transplanted, its chances of survival are low because of massive root loss.

Does it matter whether you buy or grow organic crops versus inorganically grown crops? That is a matter of opinion. Both types of crops have their advantages and disadvantages, and it is up to each consumer or grower to make their own choice.

Figure 10.7. While all plants require some amount of sunlight, every species differs in its sunlight need. Hostas are a good example of a plant that does not do well in direct sun but is perfectly happy to grow in deep forest shade.

e. wilting or washing away

When it comes to growing plants, water is often a limiting factor. If you travel to the most arid places on the planet, you will see that few plants are present. Some plants can survive in desert climates with very special adaptations but the vast majority of plants require considerable amounts of water to grow and thrive. In a natural setting, the plants that are present in any particular location have generally adapted to the amount of annual rainfall. Plants that grow in the desert do not expect to see a lot of rain. Plants that grow in jungles expect to see a lot of humidity and a lot of rain. These plants have adapted to specific regions and are not usually successful when moved out of their geographic or climactic region.

When it comes to growers and gardeners growing plants, water is a major factor in their success. Most of the agricultural and ornamental crops we grow require some irrigation, at least when first established. Small plants have weak root systems and cannot go far for water. When plants are moved to new locations, many of their roots are often left in the soil from which they came in the process of transplanting. Because their root systems are compromised, they will require more water than what can usually be supplied by rainfall. If someone is not attending to the plant everyday without a hose, it will often die.

Figure 10.8. Plants need water but too much water can be as bad as not enough water. Annual floods damage millions of acres of crops annually, all over the world.

When plants do not have enough water, they wilt. When plants wilt, they essentially become dehydrated and their cells physically collapse. Water is critical for many different plant functions, but one of its most important functions is to provide rigidity for plants. While woody plants do have the strength of wood to hold them upright, nonwoody (also called herbaceous) plants use the water in their cells to hold them erect. The water in each cell inflates the cell like a balloon. Cell walls can provide some structure, but the water inside the cell is necessary to make sure that the membranes and walls do not collapse.

Wilting can occur at different levels. When plants first begin to wilt, they can easily be rehydrated with moderate amounts of water. However, if plant cells dehydrate too much and the plant wilts too severely, no amount of water will resuscitate that plant. The term "permanent wilting point" describes the point at which there is so little water left in the soil that the plant will not recover from wilting. When a plant reaches the permanent wilting point, it becomes a dead plant. The permanent wilting point is different for every soil/plant combination. Field capacity is a term that describes the amount of water in the soil after the soil has been saturated and allowed to drain for a while. While there is a lot of water in most soils at field capacity, there is much less water at the permanent wilting point (PWP). There will still be water in the soil at the PWP but it is held so tightly in small pore spaces that the plant cannot physically suck it out of the soil.

Not all crops require irrigation. In some agricultural settings, natural precipitation may be enough to ensure a successful crop yield. However, if a long-term drought occurs, plants will not receive enough water and the crop will fail. In very large agricultural systems, the planted crop may be too large to irrigate if a drought occurs, and nothing can be done to save it. In 2012, a major drought hit the United States, and a significant amount of the Midwestern corn crop failed, costing growers and consumers hundreds of millions of dollars. There is so much corn in America and it's value is so low per acre, it doesn't make economic sense to water it. If the clouds don't water the corn, it does not get watered!

In very high-value crops or very small production fields, irrigation is often used as a backup to ensure that the crop does not fail. Most growers would prefer not to irrigate if possible. Water can be an expensive input, and the fuel costs involved in pumping water can add up quickly. A small thirty to hundred acre farm might spend $10,000 on the cost of fuel to run water pumps in a dry year! Some crops absolutely require irrigation. In some very arid desert environments, small production fields and even golf courses have been established. Because there is no regular rain in these desert environments, plants must be continually watered. If you fly over the Southwestern United States, you can often glimpse some of these cropping systems. In states like Arizona, large green circles of crops (potatoes, cotton, tomatoes, soybeans, etc.) can be seen from the air, completely surrounded by desert. This is an example of pivot irrigation. Because these environments are reliably warm and dry, plants grow very well with few diseases and pests. Growers save money by not needing pesticides. And as long as the irrigation system continues to run, the crops will thrive. And pivot irrigation tends to reduce water waste, meaning less water is required to maintain a crop than if it were planted in long rows. This technique is not limited to desert environments; many fields in the Midwest also use this approach because it relies on simple systems that are very efficient.

While too little water can be a problem for growers, too much water can be just as bad. If irrigation is available, more water can always be added to a field. But if too much water is present, nothing can be done to take it away. Heavy spring rains often cause problems for growers when they try to establish a crop. If the soil remains too wet for too long, growers cannot plant seed, and any seed that is planted may wash away or rot in the fields before it can germinate. If a crop has already been established, too much water can suffocate roots and cause plants to turn yellow. Yields may be low and plants may be susceptible to disease. Disease is a major factor when water is present, and continual rain and overcast conditions can spark major disease outbreaks that can destroy an entire year of crops. The disease called late blight is a significant concern in wet summers, destroying entire potato and tomato fields. In 2009, most of the potatoes and many of the tomatoes in New England were killed by an outbreak of late blight, the same pathogen that killed all the potatoes in the 1850's and led to the Irish Potato Famine.

Houseplants and home lawns are extremely susceptible to overwatering. Often, the most common killer of houseplants is water. Homeowners who lovingly water their plants frequently overwater them, causing roots to suffocate and disease to proliferate in the roots. After months of daily overwatering, plants collapse and roots rot away. It is often better to allow house plants to get slightly dry and even a little wilty than to overwater them. A similar situation also often occurs in home lawns, particularly when homeowners have irrigation systems. When a homeowner does not have in-ground irrigation, it becomes difficult to regularly water turf, and plants get spotty amounts of additional water. The plants may be a bit dry, but they do not usually die from drought—the turfgrasses actually have drought survival mechanisms. When a homeowner does have an irrigation system, they invariably overuse the system, drowning the turf. The homeowner kills the grass with love. While too much water is not nearly as common a problem as too little water, plants can be negatively affected by both. Water management is a critical aspect to successfully growing most agricultural or aesthetic crops in the field or the home landscape.

f. hot or cold?

In addition to water—and often in combination with water—temperature can play a major role in how well or how poorly plants grow. Different plants grow best in specific regions, and temperature is usually primarily responsible for where different types of plants grow. Plants may be divided into groups based on the time for one life cycle to be completed. A complete life cycle in this case occurs from the time a plant grows from a seed into a mature plant that produces seed for the next generation.

Annuals are able to grow from seed to maturity within one year, or growing season, and die when the winter comes. However, the lifespan of some annual plants can be extended by growing them in very warm climates or in greenhouses. Some annuals may complete growth in significantly less time than a full season. For example, as few as 45 days are necessary for some plants in the Poaceae and Brassicaceae families to go from seed to seed.

Biennials usually require two growing seasons, with a cool period in between the two warmer seasons. The cool period is essential for flower development. During the first year (season) of growth, vegetation is produced (that is, the leaves). Many biennial plants form a rosette of leaves growing close to the soil during this first stage. The second growing season (year) after flowers have been developed usually results in much taller plants that terminate in the flower or inflorescence. Occasionally, a cool period during the first season of growth may be sufficient to produce flowers on biennials. This phenomenon is termed 'bolting'.

A perennial is any plant that grows in multiple seasons and lives beyond two years. Perennial plants can survive the winter and will regrow the following year. But some plants are only perennial in warmer climates and cannot survive harsh winters. Some perennials, such as the century plant (*Agave Americana*) grow for ten to thirty years, flower once, and die.

And then there are additional considerations for different types of plants. Some plants require very warm temperatures and simply cannot complete their life cycle in cold environments. Other perennial plants require a cold winter to be successful year after year and actually depend on a period of cold weather. Some plants can grow in the wrong climate but grow much better in their preferred climate. Many plants that only grow to the height of a shrub in the Northern United States may grow to the height of a full tree farther south, where temperatures are high and the growing season is longer. Plants are dependent upon temperature for growth. The warmer the temperature, the faster they grow. The cooler the temperature, the slower they grow. People can moderate their body temperature; external temperature won't make you grow faster or slower. But plant growth is in direct response to external temperature. Warm temperatures provide heat for the plant and this heat speeds up their metabolism, allowing them to do more in less time.

The term "growing season" refers to the length of time that crops can be successfully grown in a particular region. In cold climates, the season is short. In warm climates, the season is long. In a tropical or subtropical region, the season may be all year long. Growing season is primarily affected by temperature, but other factors such as the elevation, the amount of light in a particular region (which varies throughout the year), humidity, and even rainfall can have an impact on the length of a growing season. The United States is divided into different growing regions, generally based on the dates of first and last frost. These regions can be viewed on a USDA growing zone map. The zones do not run strictly north to south—elevation has a huge impact on the length of growing seasons as can be seen on the map.

In addition to growing zones, growers and horticulturalists will also talk about growing degree days. Growing degree days (GDD) are a way to measure how much heat a plant has been exposed to. Because plants grow in relation to the heat they recieve (assuming they are in a favorable environment with enough water and sunlight), the amount of accumulated heat they receive each day can be used to predict whether they will complete their life cycle in a growing season and when they will flower. Often, horticulturalists will use GDD as a way to compare what happens from one growing season to the next. GDD can tell you whether the crops are growing faster or slower than last year and what to expect from a harvest. GDD can also be used to determine the appropriate timing to control insect pests and disease. But as important as temperature is to the growth of plants, it is one the few things that growers, farmers, horticulturalists, and gardeners have little control over in the field.

11

a. plants get sick, too

every living thing has the potential to get sick. In our everyday lives, we generally only consider the sicknesses and diseases that we, as people, experience. On occasion, we have to treat our pets for illness, and every once in a while, we may hear about a sickness affecting farm animals or wildlife. Most of the time we don't really think about plants and their diseases. As it turns out, plants are extremely susceptible to disease and battle pathogens on a daily basis. Plants are everywhere around us. Everywhere that plants exist, they are under attack.

Plant diseases do share a few similarities to human diseases, but they are generally very different. These differences manifest themselves in many ways. Firstly, plants do not have an immune system. Humans can become completely immune to many diseases, once recovery from the disease has occurred. Once a person has been infected with the herpes virus that causes chicken pox, that person will never get the disease again. However, the virus that caused the disease will remain in the body at very low levels but will be asymptomatic (not causing symptoms associated with dis ease). Humans have extremely complicated active systems for recognizing pathogens, arresting pathogen activity, and killing pathogens. Plants do not have nearly as complicated a system or set of systems. While plants can develop some resistance to disease through very specialized mechanisms, plant defense mechanisms are actually very limited. This is not to suggest that plants cannot fight disease. Plants do fight disease effectively, but once a plant is infected, the disease will often remain a part of that plant for the rest of its life. Years later, the disease may ultimately contribute to the death of the plant.

The most common reason for plant death in nature or in agriculture is disease. A maple tree may live to be 20 years old and then suddenly become infected with a pathogen called *Verticillium*. This fungus will infect through the roots and move into the vascular system of the plant and slowly kill it. In rainy years, the plant may recover and look relatively healthy. In dry years, the plant may look sicker and start to lose branches. After another 20 or 30 years, the tree will go into decline and lose most of its branches. Secondary pathogens will start to attack the tree, and eventually it will die or fall over as it rots. Trees are an excellent example of plants living with disease because they do have such extended lifespans. Few trees present in nature or in managed landscapes are free from heartrot, a condition that may or may not kill the tree, given enough time. Other plants, such as those grown in agricultural settings, may outpace disease by virtue of harvest. Harvesting a crop and destroying the remaining plants (by tilling or just by frost) may prevent disease outbreaks from ever occurring.

Similar to human diseases, there is variation in the severity of diseases that can infect plants. Some plant diseases will remain active at a low level throughout the entire life of the plant, rarely causing significant damage. Other diseases can kill a plant over time, but this may take decades. Some plant diseases can kill in a matter of weeks or days, while some diseases are just cosmetic or may be temporary. Leaf spots on a maple tree generally will not hurt the tree—they just look ugly. And that same leaf-spot disease will no longer be present after the maple tree drops its leaves in the fall. When new leaves emerge in the spring, they will be disease free!

In addition, not all plants are susceptible to the same diseases. Humans and dogs typically do not get the same diseases. Similarly, one plant species often does not acquire the diseases of other plant species—although some diseases have a broad host range and can affect (and infect) many different species. This limited ability of plant pathogens to attack many different species is often used in agriculture to prevent disease outbreaks.

Just about all plants are infected with some type of pathogen. But plants have developed mechanisms to defend against disease in some cases and simply cope with disease in other cases, living in conjunction with the pathogen indefinitely. Plants and their pathogens have evolved together for millions of years and continue to evolve. In both nature and in agriculture, plants, plant pathologists, and plant breeders are in a constant arms race with diseases. Just as we and the plants learn to cope with or control a disease, the pathogen that causes that disease will often shift its strategy, overcoming

Figure 11.1. Although mushrooms are the most commonly encountered fungi, the fungi that cause plant diseases are usually microscopic. While the symptoms are easy to see, the pathogen itself is much harder to find.

plant defenses and once again becoming a threat to plant health. In nature, those plants that have achieved success in combating disease are the plants that still survive and thrive today. Those plants that could not cope with disease became extinct hundreds, thousands, or millions of years ago.

b. the pathogens

There is a wide variety of organisms that cause plant disease. These are called pathogens, and they are usually microorganisms (very small, usually microscopic). The plant that gets sick is called the host. In order

Figure 11.2. Fungal diseases come in many forms. Rust diseases form orange pustules all over plant leaves. In this case, a rust is slowly killing a raspberry bush, leaf at a time.

for a disease to start, you need three things: a pathogen, a host that the pathogen can attack (not all pathogens and hosts are compatible), and the right environment. Plant pathogens can rarely cause disease all of the time. Usually, a pathogen can only become infective under a specific set of environmental conditions that usually includes the proper temperature and moisture levels.

The most common group of plant pathogens are the fungi. Fungi are one of the major kingdoms of life on the planet. Most fungi are not pathogens—most fungi are saprobes. A saprobe is an organism that consumes dead things. When a tree dies and falls to the forest floor, it will decompose. After a few years, there will be no trace of the tree left. Fungi are primarily responsible for decomposing these dead things. When you walk through a forest, mushrooms are abundant. These mushrooms are the most common sign of fungi, but they are usually the smallest visible part. One small mushroom may have a huge amount of fungal mycelia underneath it, living in the forest soil (the mycelium is the microscopic, tubular body of most fungi). Not all fungi are saprobes, however. When a fungus actively attacks a live plant and tries to consume that plant or kill it, that fungus is considered a pathogen as it is causing disease. Some fungi can cause disease but are less aggressive, penetrating plant cell walls with a haustorium and slowly removing nutrients from the plant's membrane. Even though these less aggressive pathogens may not kill a plant, they can still weaken it and interfere with its ability to produce seeds. Not only does this stop the plant from spreading in a natural setting, it prevents growers and farmers from harvesting a crop. Most plant diseases are caused by fungi, and most of the crops lost to disease are lost to fungi. Fungi can also cause disease of people and animals, but this is much less common. Athlete's foot is a common fungal pathogen of people, but one of just a very few.

Fungi can cause many different diseases with highly variable symptoms. Such symptoms are the result of a disease attack—that is, the things you can see the pathogen doing to the plant. The most common diseases are leaf spots, where the pathogen causes small spots on the leaves, stems, or fruit. Sometimes a spot will proceed into a blight, where all of the tissue is completely killed. Wilts are another type of fungal symptom. When a fungus enters into the xylem or phloem, water and nutrients cannot flow, and the plant will wilt. Root rots are also common, and when a fungus kills plant roots, wilting can also occur. Fungi can cause even more unusual symptoms, from scabs to strange overgrowths.

Second to the fungi in ability to cause plant disease are the bacteria. While fungi are actually very complicated organisms, bacteria are some of the simplest organisms on Earth and are very different from fungi. Fungi are composed of many cells, and even though they are microscopic, they can be quite large. Bacteria are individual cells and are always microscopic (although large colonies of individuals can be observed with the naked eye). Fungi, like plants, possess a nucleus for controlling cellular activities and storing DNA. Bacteria do not have a nucleus, and their DNA mingles with the rest of the cytoplasm. One of the most significant differences between fungal and bacterial disease is that fungi can actively penetrate into plants by punching holes into them. Most bacteria do not have this ability. In general, bacteria have to enter into plant wounds. Bacteria also often produce toxins, or poisons, to kill plant cells, a relatively uncommon feature in fungi. While many fewer bacteria cause disease on plants, some can be extremely aggressive and cause significant levels of damage and disease. Fire blight, caused by the bacterium *Erwinia amylovora*, can completely kill small apple trees and destroy a grower's harvest before it ever gets a chance to mature. In the case of this particular disease, insects spread the bacteria from tree to tree as they pollinate flowers, demonstrating a complex evolutionary adaptation of the pathogen with insects. Unfortunately, trying to manage diseases caused by bacteria is extremely difficult, often more difficult than for fungal diseases. Despite their differences, bacterial pathogens can cause all of the same symptoms as fungal pathogens, often making it difficult to determine whether a disease is caused by a fungus or a bacterium.

Figure 11.3. Pumpkins often succumb to powdery mildew in the late summer, just before they begin to ripen. The disease is identified by white fluffy patches on the leaves, caused by a fungus that can potentially destroy all of a plants leaves and reduce pumpkin size and fruit yield.

Viruses are the third group of pathogens to cause plant disease. People get viruses all the time, and some of the most common human viral diseases include the common cold, the flu, chicken pox, and even warts. Plants can also get viral disease, but viruses have a much harder time attacking plants than people. Viruses are the simplest form of life. In fact, viruses are so simple that many (or most) people do not consider them as life. Defining life can be a difficult thing—as much a philosophical task as a scientific one. There are many qualifications necessary for something to be considered "alive." One of those is that a living thing must reproduce. While viruses do reproduce, they don't do any of the other things that living organisms do. A simple virus can be nothing more than a strand of DNA or RNA. A more complex virus can be composed of multiple strands of DNA and a protein coat. However, a virus does not have any cell walls, any cytoplasm, or any organelles. Most importantly, a virus cannot reproduce outside of a host cell. Viruses take over a cell and force that host cell to make more viruses. A virus lying on a kitchen table cannot do anything but sit there. But if it wiggles into a cell, it can cause havoc. Because viruses do not last long outside of their host and have to be passed from plant host to plant host by physical contact (plant viruses are not typically airborne like a cold virus might be), they have a hard time spreading from plant to plant. Plants don't move like you or I do, and physical contact between two plants is much less likely than physical contact between people. Compared to viruses, fungi and bacteria have pretty effective mechanisms for surviving in a dormant state and can almost always spread in the wind, the water, or on insects and other animals.

Figure 11.4. Leaf spots are some of the most common disease on all plants. Eventually spots coalesce to form large blights.

The most effective way for viruses to travel from host plant to host plant is by hitching a ride with insects. Not all viruses require insects to move them around, but insects can travel far and visit many plants in a single day. If a virus can be moved around by an insect, it has a much better chance of spreading than if it required physical contact between plants. Most insect-vectored viruses (we use the term vector to describe an insect that carries a virus or other pathogen) actually get sucked into the bodies of insects that feed on plants and are spit back into a new plant when the insect finds a new food source. Another key difference between viruses, fungi, and bacteria is that viruses usually do not kill their host plant. Viruses are slow to spread. Because of this, they may never kill a host they infect, just cause small leaf spots, mosaics, and morphological oddities while they wait around for new hosts to come by. Similar to some of the less aggressive fungal and bacterial diseases, however, they can reduce yields and prevent seeds and fruit from being produced—a major concern for growers.

The final group of pathogens that are microorganisms are the nematodes. Nematodes are actually animals and are very closely related to insects but are usually microscopic. Nematodes look like very small worms and can sometimes be seen with the naked eye if they are in groups or if the nematode is a particularly large specimen. Most nematodes that attack plants live in the soil, feed on roots, and lay eggs in the soil. Most nematodes are not pathogenic, but all plant roots have at least some low level of nematode pathogens. Nematodes are similar to viruses in that they cannot attack a dead plant. A virus uses the cell's machinery to replicate, so it cannot attack a dead plant or survive in one. A nematode does not use a plant's cellular machinery but sucks the contents out of live cells. If a cell is dead, it cannot use its stylet (which is like a syringe) to remove the liquid contents of the cell because the cells dry up and wither. Similar to a virus, a nematode pathogen does not want to kill its host because it will not be able to feed. However, nematodes will often pull so many resources out of a plant that it will never produce a harvestable crop—it will be stunted, its roots will be shallow, and it may wilt frequently. In some cases, populations of nematodes can become so high that the nematodes actually kill the plants they feed on, but then the nematode populations will rapidly drop as they run out of food. In some cases, nematodes can even act as vectors and transmit viruses from one plant to another.

Not all microorganisms that interact with plants are pathogens. Many fungi, and even some bacteria, undergo complex relationships with plants. These relationships are called symbiotic, meaning that both the plant and the pathogen prosper. The relationship often resembles an infection. Fungi called mycorrhizae infect the roots of many different plants. When the fungi infect the roots, they take plant carbohydrates, similar to what a pathogen would do, but without causing damage to the plant. In return, the fungus will provide additional water, nitrogen, and phosphorous for the plant's use. Because the mycorrhizae are widely distributed throughout the soil, they can greatly increase the absorptive capacity of plant roots and help the plant be more competitive in nature. In agricultural settings, mycorrhizae can be added to the soil to try and increase crop production if the mycorrhizae are not already present.

Figure 11.5. Nematodes can cause severe damage to many plants. In this case, nematodes are damaging lettuce by galling the lettuce roots and keeping the plants from forming marketable heads.

Members of the plant family Leguminosae (also called Fabaceae) also have a complex relationship with microorganisms, specifically bacteria in the genus *Rhizobia*. In this case, the bacteria infect roots and form very localized nodules. These nodules contain colonies of the bacteria that absorb nitrogen from the air spaces in the soil and turn it into ammonium ions (NH_4^+) that the plant can absorb in a process called nitrogen fixation. While plants can absorb carbon dioxide through their leaves, they cannot absorb gaseous nitrogen, even though the air we breathe contains about 78 percent nitrogen. In exchange for the nitrogen, the plants give the *Rhizobia* carbohydrates. Not only will the nitrogen help the plants grow, but it allows the plants to grow in very poor environments where few nutrients are available in the soil. It can also reduce the level of fertilizer necessary to grow a crop and can allow crops to be grown in a "low-maintenance" fashion without too many other inputs.

Figure 11.6. Plants can also attack other plants. Dodder is a bight orange vine that clings to other plants, slowly stealing the nutrients from infected plants as it climbs along them.

c. pests and herbivores

Pathogens are different from pests. For reasons more historical than practical, the pathogens that cause disease are fungi, bacteria, viruses, and nematodes. These organisms are usually microscopic and cause what most people consider an infection; that is, they literally get inside of the host. Pests are not microscopic and do not usually get very far into their hosts. Like pathogens, pests cause damage but do it from the outside, and are not usually systemic (they do not move through the whole plant). The most commonly identified pests are insects. Insects cost growers many millions of dollars a year. Insects feed on plants in two general ways: chewing and sucking. Chewing insects (like beetles) will feed on plants as larvae (insect babies) and as adults and can cause significant damage to leaves, stems, and roots. Ladybugs are beetles that chew on other insects. Grubs are beetle larvae that chew on plant roots. Sucking insects (like bugs—yes, the word "bug" actually means something! A bug is a sucking insect belonging to the insect order Hemiptera) will use a proboscis that is similar to a syringe to poke a hole in plant cells and suck out the contents. Aphids are a very common plant pest that are sucking bugs. Other common sucking bugs are mealy bugs, scales, and thrips. Bedbugs are also sucking bugs, but they feed on humans, not plants.

Insects and nematodes are closely related and undergo very similar life cycles. They both suck out plant contents through a syringe-like mechanism, they both lay eggs, and they both molt (molting is a maturation process by which the animal grows a new, larger skin/body and sheds the old one). They are so similar in so many respects, they should probably both be considered pests. However, nematodes are traditionally considered pathogens because they are microscopic and studied by plant pathologists, while insects are macroscopic and studied by entomologists.

Weeds are another group of extremely important pests. In fact, weeds are the most important group of either pathogens or pests. While weeds do not feed on or attack plants (with a few rare exceptions), weeds steal water, nutrients, and even sunlight from crop plants. In nature, weeds do not exist. The definition of a weed is a plant that is growing in the wrong place or a place it is not wanted. In a field of corn, any other plant becomes a weed. In nature, it is difficult to determine what is a weed because none of the plants were presumably placed there intentionally. In some cases, natural weeds can be easy to identify. These natural weeds are usually invasive plants that have come from far away and began to crowd out the native population of plants. As mentioned earlier, Kudzu is a vine from China that was brought to the United States more than 130 years ago. However, it is an extremely fast-growing legume and can be difficult to

Figure 11.7. Many insect pests attack agricultural and ornamental crops. Aphids are a common plant pest that not only reduces yield but can transmit plant viruses.

control when it escapes from the areas where it was originally planted. It has continued to spread throughout the Southeastern United States and engulfs thousands of acres full of native plants every year. Even though it is present in natural settings, this plant is an invasive weed.

In agriculture, weeds can crowd out an entire field of cultivated crops in a month or two, leaving a farmer with no harvest at all. The list of weeds is extremely long, and growers spend a considerable amount of time, energy, and money combating weeds. Weeds respond to the same things that crop plants do: water, fertilizers, and excellent soils. As a consequence, anywhere that crop plants can grow successfully, so can weeds. If you walk into any lawn and garden store, you will see that the bulk of the pesticides available for crop production are herbicides designed to kill weeds.

In addition to plants competing with crops as weeds, some plants actually attack other plants. These plants can be true pathogens (they take water, nutrients, or carbohydrates from their host), or they can be hemiparasites (they take just water from their host). One of the most familiar hemiparasites is mistletoe. Mistletoe is an evergreen plant that attacks many different trees. Birds eat the mistletoe berries and when the birds poop, the berries splatter onto tree branches. The mistletoe berries will germinate and penetrate the tree branch. As the mistletoe grows, it will steal water and nutrients from the tree but produce its own sugars and carbohydrates. While a small number of mistletoe plants may not hurt a tree, a large number of them or a very large mistletoe plant can do significant damage—and even kill the tree. Mistletoe is nearly impossible to control. It will often come back if cut because it penetrates so deeply into its host plant, and it is usually high enough in a tree to make it difficult to get to. Dodder is another common plant that attacks other plants, but dodder is a true parasite. Dodder seeds germinate in the soil and produce a small orange vine that climbs onto a plant. Dodder steals water, nutrients, and carbohydrates from its host and can cover entire trees or entire fields. Dodder is also very difficult to get rid of. It must be removed by hand, or the crop must be destroyed.

Herbivores are the final group of plant pests. Although insects are sometimes placed in this group (an herbivore is anything that eats plants), the term is often used to describe other animals that feed on plants. Typically, this would include deer, rabbits, groundhogs, crows, other birds, and even sometimes coyotes. Slugs might also be commonly placed in this group. The difference between large animal herbivores and insects is not only the way they feed but the way they are controlled. Deer are a particularly difficult herbivore to contend with. In suburban settings they can destroy a home landscape entirely. In agricultural settings, they can dramatically reduce harvest crop. After a couple months of deer feeding on a ten-acre corn field, half of the corn may be gone. While weeds and insects are often dealt with using pesticides, larger herbivores are dealt with using fences or other physical barriers, offensive-tasting foliar sprays, visual deterrents, loud noises, and finally, bullets.

d. fighting disease in agriculture and nature

Plants can also produce chemicals called phytoalexins to combat infections. These chemicals are highly toxic to plant pathogens and include alkaloids, terpenoids, and other compounds. Some of these products are produced all the time; they are constitutive. Other products are induced; they are only produced when the plant is attacked. Some plants also produce these materials to prevent herbivores from eating them. Alkaloids are a common type of chemical that have often been used in human medicine. But plants often produce these materials, either randomly or as a means for deterring people and animals from eating them. Nightshade is a plant in the Solanaceous family that produces a number of different alkaloid toxins. When ingested in small amounts, some of these alkaloids can be medicinally useful. When ingested in large amounts, they can cause hallucinations or even death. Even a single leaf can be fatal.

Figure 11.8. Deer are one of the most problematic pests in the suburban and agricultural landscape, causing significant amounts of damage to crops like corn and ornamentals like ivy and hosta.

In agriculture, the hypersensitive response can be useful, but natural plant defenses are not usually strong enough to provide complete protection. The hypersensitive response is a term used to describe the phenomenon of a plant killing off cells or entire tissues to stop an invaded. When the plant detects a small infection, it immediately tries to block the spread of the infection by sacrificing those parts of itself closest to the infection. If the pathogen cannot move through the dead cells, it is contained and the infection is stopped. In addition to natural defenses, growers resort to a wide range of different methods to protect plants from diseases and pests. The most common method of protecting plants is to apply pesticides. Pesticides exist for fungi, bacteria, nematodes, insects, and weeds. If pesticides are applied preventively before the disease occurs, it will usually be prevented entirely. If the pesticide is applied after the disease begins, it can probably be stopped well enough to allow for a successful harvest. Unfortunately, pesticides are expensive and they are usually poisonous to people at some level. As a consequence, most growers limit the amount of pesticide they use to the bare minimum. This is often done through the use of integrated pest management (IPM). IPM can encompass many different technologies to fight disease and pests, but it always relies on a threshold. A threshold is the point at which a disease or pest has gotten bad enough to cause an economic loss. A small level of disease or pests is usually acceptable. The threshold is the point at which the loss in crop yield and profit is large enough to warrant chemical action. Using thresholds allows growers to minimize chemical applications as much as possible.

In addition, other techniques like plant resistance can be used to reduce disease and pests. Not all varieties of tomato get every tomato disease. If a grower knows that a certain disease is going to be a problem, that grower can find a variety that is resistant. Plant breeders spend huge amounts of time coming up with new varieties that are disease- and pest-resistant. Sometimes this is done through age-old breeding techniques; at other times, this is done through molecular engineering. Cultural practices are also extremely important in managing disease. A cultural practice is something that involves management. For example, cleaning up all the dead and rotted potatoes after a potato harvest instead of leaving them in the field or in piles to infect the next year's crop is a cultural practice called sanitation. Growing plants cleanly can dramatically reduce disease, as can starting with clean seed or any clean starting material. Sometimes, a disease can be quarantined. By keeping a disease isolated to a localized area, it can be prevented from spreading to other plants and other fields.

Unlike human and animal medicine, however, plants are disposable. It is often cheaper and more cost-effective to throw out sick plants to prevent a disease from spreading to healthy plants than to try and cure the sick plants. This process is usually called roguing. Entire plants can be removed or just sick plant parts, depending upon the severity of the disease and its ability to spread. It also may be better to destroy a whole field and start again instead of harvesting. When people or animals get sick, we often spend as much as is required to make them healthy. When plants get sick, the economics of agriculture determine the outcome of the plants in question.

e. injury

Diseases and pests are not the only things that can cause plant damage. When a plant is harmed by something like hail, wind, pollution, or even too much rain, the damage is called abiotic (meaning "without life") and is considered injury. Injury typically occurs in a very short period of time, although damage can be permanent. Injury can take many different forms and be caused by many different factors. Insects are often considered to cause injury, even though they are alive. A major difference between injury and disease is that it does not spread. Because injury is not an infection, once the cause of the injury disappears, the damage is limited to what has already been done. Treating injury is very difficult. In most cases, plants will be left to fend for themselves. Either they will recover from the injury—or they will not.

12

a. natural plant adaptations

as discussed in Chapter 6, plants have evolved rapidly over the past 100 million years (if we can call anything that happens in that amount of time "rapid"). Besides those adaptations that led to leaves, stems, roots, and flowers, other important adaptations have also evolved that have allowed plants to thrive and spread across the globe. Some adaptations are unique and exist within just one species or within a small group of related plants. Other adaptations are more widespread and can be found among many thousands of species. Some adaptations are morphological; they affect the shape and structure of a plant. Other adaptations are chemical; that is, plants can produce molecules that protect them or give them a competitive advantage. Many adaptations allow plants to survive in extreme environmental conditions and may comprise an entire suite of individual adaptations that are morphological, chemical and developmental. Not all evolutionary changes turn into useful adaptations. Some changes are negative, reducing a plant's ability to thrive and sometimes actually killing the plant. Some changes are neutral; they do not provide for a positive adaption nor do they take away from a plant's fitness. Over millions of years, however, those adaptations that allow a plant to live longer, spread farther, and produce more seeds will predominate and become a permanent part of a plant's genetics and biology.

The use of the term "chemical changes" or "chemical adaptations" can be a little confusing because everything a plant does is chemical. Plants are comprised of biochemical systems and mechanisms that drive every second of their lives. However, when we talk of chemical changes in the evolutionary sense in this chapter, were mean secondary metabolites or chemicals that a plant produces in addition to all of it's daily biochemistry. These changes are not necessary, a plant could possibly live without them. However, these changes give a plant an advantage over others.

Chemical changes are the most common type of plant evolution and the easiest for plants to undergo without also undergoing a substantial change in lifestyle or habit. Plants can produce chemicals for many different purposes. Often, these chemicals are used as a defense mechanism, protecting them from herbivores, other plants or insects, and fungi. The Australian lemon bottle brush plant (*Callistemon citrinus*) produces a chemical called leptospermone. This chemical leaches into the soil around the plant and prevents the germination of other plant seeds. It can also be herbicidal and kill established plants if produced in high enough amounts. As a result, the bottle brush plant can grow without competition from other plants. Leptospermone was chemically modified and is now produced as a highly effective synthetic herbicide. Black walnut (*Juglans nigra*) trees utilize a similar mechanism by producing a chemical known as juglone (notice the genus name of black walnut?). Not all plants are susceptible to juglone, but those that are will die quickly when planted under a black walnut tree. Both of these examples are called allelopathy: when the chemicals produced by one plant damages or otherwise affects the growth and survival of another species.

Other defensive chemicals are not intended for use against other plants but against herbivores, insects, and fungi. One of the most common compounds, already mentioned in Chapter 9, is tannin. Tannin or tannic acids are found in thousands of different plants throughout the plant kingdom. Usually these compounds are located in plant vacuoles or in plant cell walls and cuticles. Because tannins are moderately toxic, they need to be stored away from the parts of the plant cell that are metabolically active so the plant does not hurt itself with the tannins it produces. Vacuoles are large membrane bound areas in plant cells that act like storage lockers. Anything contained in a vacuole is separate from the rest of the cell and can be stored for later use or disposal. When a plant cell gets attacked or killed, the tannins spill out of the vacuole and are exposed to the pathogen and can damage or slow the pathogen. Tannins are not particularly toxic to humans and other animals. Tannins are commonly found in popular foods including wine, tea, berries, fruits, nuts, beans, and many others. In fact, it has been found that tannins can even have positive health effects and can be used medicinally for some ailments. Other secondary metabolites were also mentioned earlier including alkaloids, terpenoids, phenols, and others. While these materials do not function in primary metabolism, they have all evolved over millions of years to provide an adaptive advantage to the plants that produce them. Often these adaptations are related to defense but can also impact wound healing and other survival mechanisms.

Figure 12.1. Pea plants are an example of a species that produces tendrils to allows the plants to cling to surfaces as they climb towards the sun.

Latex is another chemical that has evolved as a defensive adaptation. Latex is a white, thick milky substance produced by plants, usually when they have been injured. Latex can be toxic and disagreeable to many insects and can even trap them when it flows out of injured leaves. Latex contains a large number of defense substances and is particularly unpleasant and bitter to the taste. It is often moved to wound sites from other areas. In addition to its defense role, latex provides an extremely important healing function. When a plant is wounded, the latex will coagulate and seal off the wounded area quickly, preventing additional infection or water loss. Latex is an additionally useful material because it is where natural rubber comes from. The fluid latex is full of small particles of rubber. This fluid can be tapped, purified, and converted into many different natural rubber products. Other plants produce gums and resins instead of latex. While latex is water resistant, gums are water soluble and typically serve as a wound-healing mechanism in the plants that produce them. Gums have been widely used in the food processing industry as emulsifiers or thickeners. Resin is another type of defense/wound-healing adaptation that is usually produced by gymnosperms (conifers). When a conifer is wounded, it will excrete thick, gooey resin to block the wound, trap insect attackers, and kill fungi. Resins are highly complex, water-insoluble mixtures that are highly antiseptic and moderately toxic to predators. Resins also have many, many commercial uses and can be highly flammable.

Simple physical adaptations are also very common. Climbing structures, protective structures, bulbs, tubers, and rhizomes are all types of anatomical changes that allow plants to thrive and outcompete their neighbors. Plants that produce tendrils for climbing can reach above the surface of the ground for access to more light and will have a significant advantage over plants that do not have this ability. Protective structures can also be a very useful. Many raspberry plants produce prickles. These prickles are sharp and slow down herbivores that would try to eat the plants' berries. However, birds can get to the berries easily, despite the prickles. While a deer would distribute the raspberry seeds throughout the forest in its dung, birds will distribute raspberry seeds over a much greater distance and often along forest edges and in meadows. Raspberries don't grow very well in forests but they do grow well on forest edges, so the prickles ensure that the birds get most of the berries and the plants' seeds spread far and wide (although deer can eat many of the leaves on a raspberry bush, the prickles will prevent complete defoliation). Other plants produce bulbs or tubers so that they can store nutrients safely in the winter and away from herbivores. The deer can provide an example here too. Onions (*Allium cepa*) produce bulbs and potatoes (*Solanum tuberosum*) produce tubers. These are morphologically different structures but they serve the same purpose. If a deer comes along to eat the onion or the potato, it can eat only the leaves. Deer have hooves and are poor diggers. As a result, the bulb and the tuber will be safe from the deer and allow the plant to continue to survive. Often, a deer-chewed plant will respond by putting up new leaves if it is early enough in the season. Later on in the fall, those underground structures will also protect the plants through the winter. Bulbs and tubers will remain alive and dormant in frozen soil, safe from ravages of the winter and foraging animals, ready to produce new leaves in the spring.

Some of the most dramatic evolutionary adaptations are those that allow plants to live in extreme and unusual environments. Because of their adaptability, plants can thrive in almost any environment, given enough time to adapt to that environment. Mangrove plants, mentioned in Chapter 8, provide an excellent example. Approximately two dozen species of mangroves populate the coastal tropical and subtropical regions around the equator. Mangroves are special because they grow right up to the ocean's edge and even into the ocean, stabilizing marshes and tidal areas. They are able to do this with the use of prop or stilt roots that grow into the mud along coastal shallows. Most plants absorb oxygen and water through their roots. However, plant roots that grow in salt water cannot absorb oxygen from that water. Mangroves all have special adaptations such as special breathing roots (pneumatophores) or lenticels/air pores in their bark that allow them to absorb oxygen. They also cannot absorb water directly into the roots because of the high salt content. Their roots are resistant to salt penetration and filter almost all the salt out of the water they take in. Extra salt that is absorbed can be excreted out of the leaves. Because mangrove roots are often completely submerged, absorbing nutrients can also be difficult. Mangroves can absorb much of their nutrients as gas in the air in order to supplement the little available in the soil. Finally, mangrove seeds float, allowing them to move far away

The Resins inside conifers is highly flammable. One of the reasons Christmas trees are a household danger is because if they do happen to catch fire, the resin inside the tree will burn fast and hot, especially if the tree is dry. Even wet conifer wood burns fairly rapidly. Artificial trees, however, are often no better, with some being just as flammable as real trees. The resin from a few conifers of the Pacific Northwest can be distilled into highly explosive gasoline, too explosive to be used in automobiles!

from the parent plant. Mangroves have evolved an entire set of adaptations to allow them to be successful in an extreme and unusual environment. Without competition from other plants, they can flourish and provide a habitat for other organisms and a method of coastal stabilization.

While not as harsh an environment as tidal marshes, lily pads or water lilies have a set of adaptations to allow them to grow in ponds and other water bodies. Water lilies grow in many different climates and have a large leaf that floats on the surface of the water. Stems attached to the leaf anchor the plant to the bottom of a pond and contain air pockets that help it float. While all leaves are coated in cuticle, the cuticle on the surface of the water lily leaf is especially important. Because the plants are at water level, the cuticle is critical to shedding water off the leaves and allowing the stomates to operate. Water lily stomates have to be on the surface of the leaf; if they were on the bottom side they would be submerged and not able to absorb oxygen. And water lilies absorb much of the nutrients through the pond water they grow in. Like mangroves, water lilies have adapted well to their particular environment.

Other plants have evolved survival strategies for environments drastically different from mangroves and water lilies, particularly those that live in arid desert climates. The prototypical cactus is a plant evolved to thrive in desert climates. However, that same cactus is so highly evolved for the desert that it can't survive very well anywhere else. Cacti have evolved sharp spines (which are actually just highly modified leaves) to protect themselves from preda-

Figure 12.2. Many plants have evolved defensive structures to protect themselves from herbivores. Thistles are beautiful and edible but produce sharp prickles that make it difficult for people to harvest the leaves or animals to eat them.

tors. Because water is such a valuable resource in the desert, a cactus would be quickly devoured by birds, insects, and mammals if it did not utilize protective spines. Because water is so scarce, the entire stem of the cactus has evolved into a large, bulky water-storage structure. In the absence of traditional leaves, the cactus uses chlorophyll in the stem to produce carbohydrates. Cactus root systems are surprising shallow. This makes sense because when water does fall in the desert, it never moves very deeply. As a consequence, cactus roots are shallow but very widely distributed to allow the plants to absorb as much water as possible when it does indeed fall. Cacti are also very unusual because they grow extremely slowly (to conserve water) and their stomates open at night, the reverse of most other plants, which in combination with their unique metabolism allows them to conserve even more water. Finally cacti can lose more than half of their total water content before they are severely injured and they seal their wounds quickly to prevent even more water loss.

Other plants avoid extreme conditions by growing for only a short period of time. Ephemerals are plants that germinate from seed, grow, and reproduce during a short period of time when the environment is conducive to their growth. After seeding, the plants die and their seeds remain dormant until the next year in which the conditions are good for plant growth. Some ephemerals grow during the spring, typically small herbaceous plants, when it is usually wet and before the trees have leafed out and cut off their light. In the desert, ephemerals will pop up when rain falls but die back during the much longer dry periods. And some ephemerals may be agricultural weeds that start growing when a field has been cultivated and they have been pulled to the surface of the soil by a plow. The ephemerals sacrifice a long life for the sake of their offspring; they focus all their energy on producing as many seeds as possible, a strategy called avoidance.

Plants can also grow in arctic environments and a couple plants even exist in Antarctica. Plants that grow in these extremely cold environments share many of the same survival strategies: they are typically very small and low to the ground where they are not subject to high winds and desiccation, they are often covered in hairs or thick cuticle to conserve heat, they can photosynthesize at very low temperatures, they can grow under snow, they grow quickly in the polar summers and they are perennial, spending less energy on seed production and more energy on year-to-year survival.

Although Antarctica is now a harsh and barren land, mostly covered in snow and ice, hundreds of millions of years ago it was covered with life. Over millions of years the continent has moved southward and the remnants of its journey can be seen in the many thousands of plant and animal fossils recovered there.

b. breeding plants for agriculture

Plants in natural environments can evolve many different types of adaptations that will allow them to not only survive but also thrive, regardless of the environment. However, these adaptations result in plants that are very poorly suited for the garden or large-scale agriculture. There are many reasons for this, but one good example is the spread of plant seeds. In a natural environment, plants generally want to spread their seeds as far as possible and produce as many seeds as possible. This is problematic in agriculture when the harvested crops we consume are the plants' seeds. In an agricultural setting we do not want the plants to spread their seeds. In fact, we want them to hold onto those seeds for as long as possible so that we can harvest them, presumably all at the same time. For this reason and others, wild plants make very poor crop plants.

All plant species started out as wild plant species. As societies developed and matured, they took wild plants and turned them into useful agricultural plants. This process is known as domestication. Just as house cats and pet dogs have been domesticated, plants have also been domesticated. The steps that lead to plant domestication are well known and are typically very similar for every plant that is ultimately domesticated.

The first step in plant domestication is to breed plants that do not disperse their seed easily, as mentioned above, so that seeds can be harvested and do not just go back into the soil for the next generation. But plant domestication does not occur overnight. Just as is the case with natural evolution, it takes many plant generations for populations to go from shattering seed (seeds that disperse easily) to non-shattering. And because the plants would prefer to stick with their natural adaptations, it requires considerable manipulation of plant populations to produce domesticated plants. In the case of rice, a single gene controls whether plants shatter seeds or not. But early peoples who collected, saved, and replanted these seeds needed to find the plants that contained the gene mutation or mutations that prevented seed shattering. This was done through careful observation and selection of the appropriate plants. While scientists and researchers in the United States and other first world nations have been dramatically improving and engineering plant varieties substantially over the past hundred years, indigenous populations began the process thousands of years ago through domestication. The originally domesticated plant populations are usually called landraces.

The next step in plant domestication is the transformation from a hard seed coat to a digestible one. Plants evolve hard seed coats in nature to prevent their seeds from dying or being destroyed before they can germinate and produce new plants. But hard coats cannot be digested. When plants start producing soft, digestible seed coats, they can be planted as useful agricultural crops. And early peoples looked for those mutants and variants as they sifted and selected the next year of seeds to plant. Some crops can be useful with hard seed coats such as nuts, particularly those grown on trees. However, the nuts must often be shelled before sale, increasing costs, and the trees are grown in much smaller acreage because the demand for these crops is low, compared to other crops. Field crops such as corn, wheat, and soybeans will be grown over millions of acres and their economic value is tied to the large amounts of cheaply grown nutrition they produce.

Figure 12.3. Mangroves have unusual adaptations that allow them to grow in salt water and increase the length of coastlines by seeding new islands just offshore that will ultimately merge with existing landmasses.

The last critical piece of plant domestication is to alter the plants' dormancy patterns. Plants in natural populations go dormant based on environmental conditions, temperature, and light. In agricultural systems, however, plants need to germinate shortly after planting and they must germinate uniformly. If seeds remain dormant or if they take months or weeks to germinate, a grower will never harvest an appreciable yield and the crop may ultimately be wasted. In addition to breeding dormancy out of the plants or selecting for plants without a dormancy mechanism (or an altered dormancy mechanism), plants need to be harvested at the same time; therefore plants and plant populations need to be selected that

Many gardeners now grow plants from "heirloom seed". Heirloom varieties are usually older varieties, having been grown widely hundreds of years ago. Such varieties are not typically useful in modern agriculture because of some deficit in their agronomic characteristics (they may ripen too fast, can't be shipped, hard to harvest, low seed germination rate, wrong color fruit, etc.). Despite their lack of large-scale agricultural suitability, these varieties are usually very interesting and produce unique edible crops with unusual flavors, textures and colors.

Figure 12.4. Heirloom varieties are very common among gardeners and small growers. Lemon cucumbers are an heirloom type variety from the 1800s that looks like a lemon but tastes like a very sweet cucumber.

will be harvestable simultaneously. If plants mature asynchronously (not at the same time) the plants cannot be harvested at the same time and once again, the crop will be lost. This may actually be less of an issue for native populations or indigenous farmers who harvest their crops regularly throughout the season. However, large-scale industrial agricultural operations use tractors, combines, and other machines that can harvest the crop only once. Once a machine has gone through a field harvesting, there is rarely much left of the plants that were harvested from. Most industrial agriculture allows for a single harvest (although exceptions, such as pima cotton, do exist).

There are many other domestication characteristics that may be required for each different species and these characteristics can be unique to a particular species. A common problem with wheat and barley is that plants will often lodge. Lodging is when stems fall over in the field. This can be a significant problem because plants that lodge cannot be harvested by machine. While this trait can be diminished to some degree through breeding, it can still persist in plant populations and may require special management practices to minimize its impact on crop yield. This particular trait is not essential in the domestication process, but it does certainly make it easier for growers to harvest yields.

Plants are continually bred to improve their qualities. Once domesticated plant populations exist, breeders undertake plant improvement programs. Just because a plant can be grown for agricultural purposes, that does not mean it cannot be improved. Breeders are constantly making improvements, adjustments, and tweaks to plant populations in order to maximize crop yields and decrease growing costs. One of the most common plant improvements is to select and breed for plants that respond well to high rates of fertilizer inputs. Before the advent of synthetic fertilizer, it was difficult to fertilize crops. Organic sources needed to be used and without the machinery to apply the fertilizer, tons and tons of material would need to be spread by hand. This is often impractical but even when it does occur, it can be messy. Many dairy farms in New England recycle cow waste for fertilizer on corn fields. The

Fertilizer is not the only way to add nutrition to the soil. Many growers will plant cover crops to increase soil nitrogen levels. In chapter 11, the Leguminosae was introduced, a family of plants that are able to get nitrogen from bacteria attached to their roots. If a grower plants these types of plants in between the regular crop, additional nitrogen will be left behind when the crop is harvested or incorporated. Cover cropping can also reduce pests and diseases and reduce soil erosion.

corn is then fed to the cows, completing the cycle. However, when the cow waste is spread on a field, the field may stink for weeks as the manure is incorporated into the soil and decomposes. And applying the material can be a health hazard to dairy farmers and an additional expense in labor. But once synthetic fertilizers became available in the early 1900's (and widely available in the 1940's), in addition to the machinery to apply the fertilizers, breeders could now select for plants that were able to utilize the new fertilizers most effectively. Many other types of plant improvements have been made to increase crop quality, including disease and pest resistance, drought tolerance, larger seed/fruit size, more seed/fruit per plant, higher nutrition content, and other agricultural traits.

Once plants have been domesticated, they usually can no longer survive in nature. The breeding practices that we employ to make plants perform better in our agricultural fields prevent them from surviving in nature. If dormancy has been bred out of a plant, domesticated seeds that disperse in nature may germinate immediately and be killed by winter frost. Plants that produce soft seed coats may freeze or be eaten by animals before they ever germinate. And plants that are dependent upon large amounts of water and fertilizer may never produce any crop if they are left to their own devices in nature. Without the help of the humans who bred the plant, it is unlikely it will survive in the wild.

Despite the fact that people domesticate and breed plants, people ultimately become dependent upon the plants that they have domesticated. Plants domesticate people as much as people domesticate plants. Our civilizations depend on the crops we produce for our food, our fuel, our houses and structures, and our clothes. As plants were domesticated and agriculture took hold in early civilizations, populations of people went from being hunters and gatherers to being farmers. That farming was dependent upon the domesticated plants those people developed. People settled in areas where the farming was best and the soil produced the highest yields. Even today, the remnants and artifacts of native populations in the Eastern United States are most prevalent in the soils and sites best suited for agriculture. Populations of people were forever changed and irreversibly so, but so were the plants they had domesticated.

c. centers of diversity

Plants grow in many different environments, but every species or genus of plant has a location where it originated. From that location, the plant evolved all of the adaptations that allowed it to be successful. That location is often referred to as the Center of Diversity of the Center of Origin. The term "diversity" is often used because plants that have developed in those particular areas have many relatives and closely related species that have also evolved in that area. There is a large pool of closely related species containing many of the same adaptations and strengths. In fact, after a plant has been domesticated for even thousands of years, wild and weedy plant relatives will continue to grow in these centers. Scientists will often travel to these regions to collect these plants and use their germplasm (genetic material) for improving the quality of agricultural crops. Plant breeders will collect related plants and seeds and cross those wild and weedy plants into their breeding programs, incorporating traits like disease and insect resistance, drought resistance, and other useful characteristics.

Most of the Centers of Origin exist in tropical or subtropical regions. Because of the mild winters and generally good growing conditions in these environments, many of the crops that were bred for agricultural use were able to thrive and evolve rapidly. Many of these centers were discovered by the famous Russian plant breeder, Dr. Nikolai Ivanovich Vavilov during the first half of the 20th century. He traveled the world collecting seeds and plant materials and sending them back to Russia to be stored and used in plant improvement breeding programs. The primary centers of origin include Mexico and Central America (corn, beans, cotton), South America (potatoes, tomatoes, pima cotton, cocoa, tobacco, peanut, and pineapple), the Mediterranean (peas, olives, beets, cabbage, and lettuce), the Middle East and Northern Africa (wheat, barley, millet, coffee, okra), Central Asia (beans, hemp, cotton, onion, garlic, apple, pear), India and Indonesia (rice, peas, beans, eggplant, cucumber, orange, sugar cane, pepper, coconut, banana), and China (soybean, radish, peach, cherry, walnut). Many other crops exist beyond the ones mentioned here and most of them come from one of the Centers of Origins listed. Even today large institutes and research endeavors exist that are constantly collecting and improving plant germplasm from material discovered from these special places.

Not all people agree on how the different Centers of Origin should necessarily be grouped together, but that doesn't change the fact that these areas are the most important locations on the planet for plant diversity. Unfortunately, as human populations increase and encroach into these areas, the diversity of the plant material is degraded. People damage habitats, build houses, and plow up native areas for their own farms. When this happens, fewer wild and weedy plant relatives are available for future use and breeding efforts. Most people are aware of similar threats to rainforests and jungles, but many people do not realize how important other areas, perhaps areas that just look like weedy meadows, are to the continued development of important plant varieties.

And then there is global warming. Climate scientists expect that global temperatures will rise dramatically in the next century. While this will cause significant sea-level rises, it will also change weather patterns across the globe. In some regions, the prospects for growing crops may actually improve. Other regions will see extreme droughts, high temperatures, and may no longer be able to sustain much agricultural production. However, it may also negatively impact Centers of Diversity. In combination with encroaching human populations, global change may further damage or destroy these critical areas of plant diversity.

a. what is a fruit?

definitions for the word "fruit" vary widely. In a technical and botanical sense, a fruit is a seed-containing part of the plant that arises from the ovaries of flowers. Some fruit are edible and some are not. Most of the things we think of as fruit are sweet and sugary but not all fruits are this way. When you walk into the grocery store, you'll find many fruits in the produce section. Obviously, apples and pears are a typical sweet fruit that most people are familiar with. But there are many other fruits that the grocery store considers vegetables, including squash, tomatoes, peppers, and eggplants that are all technically fruits. Many vegetables are exactly that: they are botanically a type of vegetable. Celery, for example, is a vegetable because it is a vegetative part—it has nothing to do with the flowers. So why are they called vegetables when they are actually fruit? The answer has to do with taste, nutrition, and cooking uses. If we compare apples to pumpkins, we can see the apparent differences. Anyone who has ever eaten an apple knows that they are sweet. Pumpkins are much less sweet and very few people would recommend eating them raw. Most of the grocery store fruits fall into the sweet category. Nutritionally, what the grocery store considers fruits are very similar to each other, as are the vegetables similar to other vegetables. Fruits tend to have very high sugar content and many of them are very high in vitamin C. Vegetables on the other hand are usually much lower in sugar content and often have very high levels of iron, calcium, and other minerals and nutrients. Finally, cooking use also plays a role. The grocery store fruits are often eaten fresh and when they are cooked or prepared, they are used in very different dishes than the vegetables. Salads are one example. While apples occasionally find their way into a green salad, the vast majority of a green salad is composed of vegetables. Desserts are another example. The grocery store fruits are a common dessert component, but not the vegetables.

But it gets even more complicated than just taste, nutrition, and use. Different plants grow in different ways. Perennial plants grow over many years and continue to produce fruit every year. Annual plants grow for only a single season, produce their fruit, and then die. For our purposes, plants in the fruit category are plants grown commercially as perennials whose ovary and/or surrounding structures are consumed by humans. By this definition, the trees and shrubs of which we eat the fruit would fit the category. This would include apples, oranges, and even the nut crops, such as cashew, walnut, pecan, and pistachio. Strawberries, being perennial, would also be considered fruits by this definition. Melons, squash, tomatoes, peppers, and eggplants are grown as annuals in field production, which would eliminate their classification as fruits using our definition. But they are still technically fruits!

b. introduction to the rose family

Flower morphology is one of the key characteristics used to group plants in one classification versus another. There is significant similarity between flowers of fruits placed within the rose family (called the Rosaceae). The flowers tend to have five petals, ranging in color from white to pink-tinted, they are often small and delicate (but not always), and can be produced singly or in small clusters. Fruits in this family include apples, pears, peaches, almonds, plums, apricots, cherries, strawberries, and brambles (blackberries and raspberries). And obviously, roses are also a member of this family.

Figure 13.1. The rosaceous family contains many different species but it gets its name from the rose.

The family is further divided into subfamilies. Apples (*Maluus domestica*) and pears (*Pyrus communis*) are in the Pomoideae subfamily and are also called the pome fruits. The Prunoideae subfamily includes the stone fruits, also called the drupe fruits. Peaches, almonds, plums, apricots, and cherries have centers we usually refer to as pits, which are actually the drupes of the fruits or the "stone," which is extremely hard. The Prunoideae fruits are all in the same genus, *Prunus*, which is divided into subgenera. The subgenus Amygdalus includes the peach (*Prunus persica*) and almond (*Prunus dulcis*). Plums (*Prunus* spp.) and apricots (*Prunus armeniaca*) are in the Prunophora subgenus. Two plum species are popular: European plum (*Prunus domestica*) and Japanese plum (*Prunus salicina*). Cherries, in the Cerasus subgenus, include two species the sweet cherry (*Prunus avium*) and the sour cherry (*Prunus cerasus*).

Strawberries and brambles are classified in the Rosoideae subfamily. The strawberry, which resulted from hybridization among species, is *Fragaria* x *ananassa*. Blackberry is a common name for a number of species in the Rubus genus. Red raspberry is *Rubus idaeus*, while black raspberry is *Rubus occidentalis*.

Fruits that develop on trees may form on one-year-old branches, which are termed "laterals." To encourage increased fruiting, yearly pruning is necessary. Consequently, orchard owners are careful and diligent pruners, shaping and managing their trees to produce as much fruit as possible. Other fruits develop on extremely short branches consisting of wood that is two or more years old. These branches, which may grow only approximately one fourth of an inch per year, are known as spurs.

Some trees may produce fruits on both lateral and spur wood, but the majority of fruits are usually borne on one of those types. Any pruning of trees, such as apples, that bear fruits primarily on spurs must be done judiciously to avoid damaging the spurs. Peaches develop primarily on lateral wood.

To ensure the continuity of fruit among trees of the same cultivar, grafting is typically done. In addition to making the fruits genetically identical, some grafting produces resistance to diseases or nematodes, enhanced tolerance to various environmental stresses, or reduced time to maturation of the trees. In some fruits, grafting results in reduced plant growth (dwarfing).

c. apples and pears

Apples, pears, and their less-common relatives quince (*Cydonia oblonga*), mayhaw (*Crataegus* spp.), loquat (*Eriobotrya* spp.), and juneberry (*Amelanchier* spp.) contain central cores consisting of seeds surrounded by the ovary in which they developed. The fleshy tissue of each fruit, the part that we eat, is actually receptacle tissue to which flower parts were previously attached. Fruits arise from the female sexual part of each flower, the ovary.

Most fruits will not develop unless they are fertilized. Male pollen from a different plant (a plant in the same species, just not the same genus) will fertilize the ovaries and begin seed development. The developing seeds promote the formation of the receptacle tissue, which is the part of the apple we most enjoy eating. If any of the seeds don't become fertilized, the unfertilized seeds will not stimulate this tissue production, resulting in lopsided fruit that may be less marketable to consumers who prefer uniform, fully developed fruit.

Apple cultivars have been developed through extensive breeding programs, as well as being found as mutant branches on established trees. The mutant branches, which are then vegetatively propagated to maintain the clonal nature of the cultivar, are referred to as "sports." Vegetative propagation requires the branches to be cut off and attached to the roots (rootstocks) of other plants that have no other branches, in a process called grafting. Most apple cultivars are grafted onto rootstocks as a means of clonally propagating the cultivar. Several series of apple rootstocks have been developed to produce dwarf trees of varied heights. Dwarf trees are desirable because they develop fruit earlier than full-sized trees, they occupy less space than full-sized trees, and they are easier to harvest because they are so short.

Figure 13.2. Apples were one of the most important crops planted in North America as the continent was settled and continue to be an important part of the American culture.

In order to get apples, apple flowers need to be fertilized. This is a sexual process. But when new apple trees are produced, grafting is used. Grafting is a clonal or non-sexual process. Because the apples have been fertilized, they contain viable seeds that could be used to produce new trees. However, this is never done. The reason this is never done is that the apple seeds have been sexually produced and are genetically very different from the parent tree, just like every child is very different from their parents. The seeds can produce new trees, but all the characteristics of the parents will have been scrambled and the new trees may produce very different fruit from the parents. In addition, growing a tree from seed will mean it will not be attached to a rootstock, which can force the plant to become a dwarf or impart important disease or insect resistance.

The United States, New Zealand, and Japan have developed numerous apple varieties popular in today's markets. The term "variety" means the same thing as "cultivar," which is an abbreviation for "cultivated variety."

To properly flower, apples, like many temperate fruits, require a specified length of chilling temperatures, followed by adequate temperatures known as growing degree days. The variability among newer apple cultivars for these requirements allows some to be grown further south in the United States where they were previously less adapted. The most commonly grown apple varieties still require significant amounts of chilling temperatures and do best in places like the Northwest and Northeast United States. Washington State and New York State produce the vast majority of apples in the United States.

Figure 13.3. Despite the difference in appearance and taste, pears and apples are closely related. The flowers of the two species are almost identical and both require pollination by bees to bear fruit.

The predominance of apple production in the United States would give one the impression this fruit is native to our country. However, apples actually originated in the eastern part of Turkey and Kazakhstan. Apples have been domesticated for thousands of years, with the first historical record of apples appearing in the Middle East in 6500 BC. The *Malus domestica* species is originally developed from dozens of other species, which have been bred and hybridized over thousands of years. In the United States, apples were one of the most important crops of the early settlers. Every farm had an orchard and apples provided a major source of easily storable food that could be eaten throughout the winter. As settlers moved across the continent, they brought apples with them.

Pears are native to a milder climate than apples. Mediterranean areas provide suitable climates for extensive pear cultivation, but they also do well in the same areas that apples are produced. Pears have also been cultivated, bred, and consumed by humans for many thousands of years, dating as far back as ancient Rome and earlier. Pears originated in China, close to where apples originated, and have been cultivated by the Chinese for at least 3,000 years.

Pears are divided into two groups: the European pear (*Pyrus communis*) and the Asian pear (*Pyrus pyrifolia*). The European pears are most commonly found in American markets. Two of the more popular cultivars are 'Bartlett' and 'Bosc'. The 'Bartlett' pear has the shape most familiar to pears, with a top that is more slender than the rounded bottom. 'Bosc' pears are squatter in size and more oval in shape compared to the 'Bartlett' pears.

d. peaches and almonds

As noted earlier, peaches (*Prunus persica*) and almonds (*Prunus dulcis*) are classified in the Prunoideae subfamily and *Prunus* genus, along with plums and apricots. Peaches and almonds, however, are more closely related to each other than they are to plums and apricots, so they have been separated into the subgenus Amygdalus. Although the peaches and almonds you buy in the store have no outward similarities, when you remove the seeds from their pits, they are remarkably similar.

Peach cultivars fall into one of three groups. The first group may be somewhat surprising because it is the nectarines, which are simply fuzzless peaches and not a separate plant species. The remaining two groups of fuzzy peaches are distinguished by whether their soft tissue adheres to the pit or is readily separated from the pit. Clingstone peaches tend to be softer when ripe and their tissue "clings" to the pit when you get to the center of the sweet, juicy flesh. These peaches are sold primarily for fresh consumption.

Peaches whose flesh readily separates from their pits are freestone peaches. They tend to be firmer and retain their color better than clingstone peaches, which makes them more desirable for the processing industry. Processed peaches are the kind you find in cans. Clingstone peaches would be less colorful if they were canned and the cut pieces would lose their shape, becoming a soupy mush. Most peaches are a golden yellow with a shading of red at the top of the fruit. This red color is known as the blush. More recently, white peaches have become available, although yellow peaches remain the most popular.

Although Georgia is synonymous with peaches, California has the greatest production in the United States. All clingstones and most nectarines are produced in California, while freestone production is spread across much of the lower 48 states. As with pears, peaches were originally domesticated in China almost 4,000 years ago. China still leads all countries in production of this fruit.

Successful peach production is a little more difficult than apple or pear production. As much as 95 percent of the flowers are removed from peach trees in order to obtain larger fruit. Removal is usually done by hand

Figure 13.4. Although Georgia is considered the Peach State, most American peaches come from California and the species is actually native to China.

to ensure removal of flowers at proper intervals to prevent fruits from touching during development. If too many flowers are left on a tree, the tree will put its resources into producing as many peaches as possible. They will all develop to some degree, but they will be small and not very edible. The peach tree doesn't really care because the whole point of a peach is to produce a new peach tree, not feed humans. But humans, in an attempt to force the plant into producing the best peaches, limit the number of peaches each tree can produce!

Among the different types of fruit trees, peach trees are highly susceptible to poor environmental factors that may shorten their lives. Affected trees that appear healthy in the fall may suddenly die the following spring. While the pathology is not completely understood, nematodes are believed to play a major role. This phenomenon is termed Peach Tree Short Life, or PTSL. Orchards should not be replanted to peach trees where PTSL occurs. Apples can also be subject to nematodes attacking the roots in a disease commonly called Apple Replant Disease.

Figure 13.5. Almonds and peaches look very different but belong to the same genus and actually look very similar when the fruit of both trees are young.

Color is the best indicator of peach ripeness. For shipping, peaches are harvested while still quite firm to reduce bruising, but they do not approach the sweetness and juiciness of fully ripe peaches picked directly from the tree.

As mentioned previously, peaches and almonds are very similar to each other, despite their obvious physical differences. The fruit somewhat resembles a peach while it is maturing, although it is much smaller than commercially produced peaches, since flowers on almond trees are seldom pruned. Peach fruits would be as small as almond fruits if they were not as severely pruned for production. The fleshy part of the almond is called the hull. It must dry and split before the almonds are harvested. The shell of the almond is the pit. When the pit is cracked, the almond, or seed, is released.

While apples, peaches, and many pear varieties are produced through grafting, almonds can be readily produced from seed. Because of this, it is likely that it was one of the earliest tree fruits to be domesticated at least 6,000 years ago in Armenia. Unfortunately, wild almonds are highly toxic, producing lethal levels of cyanide, and domestication of the plant did require early farmers to identify and/or breed new varieties that were not poisonous. Although native to the Middle East, the greatest almond production in the world occurs in California. Just like with the other tree fruits we have mentioned, almonds need to be fertilized to produce a viable seed. With all tree fruits, bees are the most common pollinator. In order for the California almond crop to be successful, hundreds of millions of bees managed in commercial hives have to be trucked into California every February.

The navel orangeworm (*Amyelois transitella*) can be a serious pest of almonds if almonds are not harvested promptly when ripe, as well as during storage. The insect will bore into the almond during its larval stage (a small caterpillar) and eat the seed, leaving nothing behind to be harvested. Interestingly, cyanide-containing plants are often toxic to insects and it is probable that the cyanide in ornamental almonds and the first original species prevented insect damage. Fumigation destroys the naval orangeworm when it is present during storage.

Harvesting the nuts by hand would be too time-consuming, so hand-knocking of the trunks is done to younger trees, while more mature trees are machine-shaken. The nuts are allowed to lie on the ground to dry for one to two weeks, at which time they are swept up for further processing, including additional drying.

e. plums and apricots

Plums (*Prunus domestica* and *P. salicina*) and apricots (*Prunus armeniaca*) are in the same subfamily and genus as peaches and almonds (and are drupe fruits), but are classified into their own subgenus, Prunophora. Plum and apricot pits are much smaller than pits in peaches and almonds.

Plums and apricots are so closely related, they readily hybridize. That is, they can breed with each other easily. Crossing a plum with an apricot produces offspring with half their genetic material from the plum and half their genetic material from the apricot. This hybrid is named a plumcot. If a plumcot is crossed with a plum, the hybrid is called a pluot. Seventy-five percent of the genetic material is from the plum genus, while only 25 percent is apricot genetic material. The pluot is the most commonly found hybrid for sale in markets. Crossing a plumcot with an apricot produces an aprium. This hybrid is 75 percent apricot and 25 percent plum genetics. Because these unusual fruits contain varying genetic material from the two original species, they also contain varying amounts of expressed traits such as color, texture, taste, etc.

European plums (*P. domestica*) include the types most commonly used to produce prunes. Many producers now market their dehydrated fruit as dried plums, presumably to avoid any negative associations people may have with the prune label. For fresh eating, Japanese plums (*P. salicina*) are most popular. Although called Japanese plums, this species is actually native to China, much like many of the other tree fruit crops, domesticated thousands of years ago.

These two major species of plums have several distinguishing botanical features. Japanese plums have rougher bark, more persistent spurs, and more numerous flowers compared to European plums. In order to achieve proper size of the Japanese plum fruits, flower removal is necessary, similar to peaches. The lesser-flowering European plums typically do not require as much, if any, pruning. Fruit shape also differs between the two species. Japanese plums tend to be more round, while European plums tend to be more oval.

Japanese plums tend to bear fruit at a younger age than European plums. Cultivars of Japanese plum also tend to be more vigorous and disease-resistant than cultivars of European plums. Cold hardiness is greater in European plums. Japanese plums thrive best in a Mediterranean climate.

Apricots are consumed in far smaller quantities than many other fruits in the rose family. Most apricot fruits are dried, since fresh fruits spoil fairly rapidly. Dried apricots are more flavorful than fresh apricots because they are harvested later when more sugars have accumulated in the fruit. Although similar in color to peaches, apricots have a distinct linear indentation, known as a suture, along the length of the fruit.

f. strawberries

Strawberries are one the favorite American fruits of early summer and are known for their red, sweet juiciness. Unlike the previously mentioned fruits in this chapter, strawberries do not grow on trees but in small bushy clumps on the ground, usually less than a foot tall. In addition, the previously mentioned fruits all contain one of a few seeds in the middle of the fruit. Strawberries contain many seeds dispersed throughout the outer edge of the fruit and are called an aggregate fruit. An additional difference is the origin of the strawberry. While apples, pears, apricots, plums, and almonds were all domesticated thousands of years ago in either China or the Middle East, strawberries are a cross between wild North American and wild Chilean strawberries, originally bred in the 1700's, creating the hybrid *Fragaria* x *ananassa*. Wild strawberries had been grown prior to the new hybrid, but none of the wild varieties produced very large fruit or were particularly reliable. Today, California leads the world in total strawberry production.

Figure 13.6. Unlike most fruits we call berries, strawberries grow on short plants that stay low to the ground and spread from year to year by runners.

Commercial production often involves annual reestablishment, despite this species being perennial. As strawberry plants get older, they may produce fewer fruit in subsequent years, become susceptible to insects and diseases, and generally lose vigor. Replacement with new plants each year allows for higher production. Although strawberries can be produced from seed, a similar problem arises as with apples. That is, the genetics of the seeded plants will be different from the parents. But strawberries naturally produce runners (above-ground stems that travel along the soil surface) and at the end of each runner is a brand new baby plant (or plantlet). As a consequence, it's relatively easy to produce new plants—just snip the runner off the original plant and place the plantlet in the new location!

While weed-free straw mulch was commonly used in the past and is still in use in home gardens, commercial production now includes the use of plastic row covering, with slits for plants to be inserted. The plastic helps reduce weed populations, increases the rate of soil warming in the spring to hasten plant growth, and keeps the fruit cleaner.

Home garden production may involve raised beds where plants are elevated in benches. This makes the fruit easier to harvest and provides good drainage to decrease disease incidence and prevent the fruit from rotting.

Strawberries have typically been placed into two distinct groups: everbearing strawberries and single-bearing (June-bearing) strawberries. Everbearing strawberries are exactly that: ever bearing. These strawberries typically produce fruit all season long. Single-bearing or June-bearing strawberries produce a single major crop of fruit in June and then produce very few, if any, berries for the rest of the season.

g. brambles—blackberries and raspberries

Fruit produced on stems growing directly from the soil are called brambles. These stems are properly called canes. In the rose family, the two most common brambles are blackberries and raspberries. The ease of

Figure 13.7. Russia and Poland are the largest worldwide producer of raspberries but the United States also produces a significant amount of the fruits which can even come in a yellow variety.

hybridization of these and related species has increased the variability of fruits from brambles, but also blurs the distinctions among species. There are many different species and subspecies that go by the blackberry name, but the common blackberry is *Rubus fruticosus*. Red raspberry is *Rubus idaeus*, while black raspberry is *Rubus occidentalis*. Brambles are like strawberries in that they are an aggregate fruit.

The main distinguishing features of blackberries and raspberries are the appearance and harvest of the fruits. Blackberries are shiny and hairless, while raspberries are dull and hairy. When ripe, blackberries do not separate from their core, which is the flower's receptacle tissue. As the blackberries are pulled from their canes, the core is removed along with the fruit. Raspberries detach from the core, which remains attached to the cane.

Prickles of varying length and sharpness line the canes of most brambles. Although many people may consider these thorns, they are actually called prickles and are botanically different from a thorn. Roses also have prickles, not thorns. Despite this technicality, thornless (or "prickle-less") blackberry cultivars are available.

Most bramble fruits are harvested by hand, since they are easily crushed. Mechanical harvesting is sometimes done when the fruit will be processed. The short storage life of brambles necessitates rapid processing and high consumer cost for fresh fruit. Although raspberry fruits are most commonly red, there are black raspberries. Like the red raspberries, the black raspberries have dull, hairy fruit that separates from the core when the fruit is mature.

Bramble canes live for two years. During the first year, the cane is known as a primocane. For most brambles, flowering and fruiting occurs during the second year of growth. The second-year canes are called floricanes. Once the canes have produced fruit, they will die, but the plants produce new canes from adventitious buds on their roots. One group of raspberries produces fruit twice during its life cycle. The first fruit matures at the end of the first growing season, while the second fruit crop is produced the following summer.

Blackberries and raspberries have a wide range of areas where they can be successfully cultivated. The majority of the *Rubrus* production in the world occurs in Mexico while the majority of the production in the United States occurs in the Pacific Northwest, two drastically different climates. And blackberries are commonly found wild throughout the world, growing on the edges of forests and waste places.

a. blueberries

although the rose family contains many of the common fruit crops, there are a wide variety of other fruit crops that belong to completely different families. Blueberries are an excellent example. Blueberries (*Vaccinium* spp.) are native to the Eastern United States and northward into Canada. Individual species and species hybrids are now commercially produced in much of the United States and in other countries, including Poland and the Netherlands. These fruits are in the Ericaceae family, which also include plants grown for their flowers, such as rhododendrons (*Rhododendron* spp.), azaleas (*Azalea* spp.), and mountain laurel (*Kalmia* spp).

While many of the fruits we harvest are produced on trees, blueberries grow on shrubs. Individual stems of the shrub grow for about four years and new stems eventually replace older stems. The growing point, or crown, for each stem is compressed and located near the ground. Stems may also grow adventitiously from the roots. Some blueberry species are adapted to cooler climates of the northern United States, while others do better in the southern United States. Northern plants require longer hours of chilling temperatures during the winter months in order to flower and produce fruits compared to southern plants.

V. corymbosum, the northern highbush blueberry, may grow upward as much as 12 feet, but is usually maintained at shorter heights in commercial production for easier harvesting. Lowbush blueberries, *V. angustifolium* and *V. myrtilloides*, have not undergone extensive breeding and genetic improvement and are typically called wild blueberries. Some farms are simply managed native plants that usually stay less than two feet tall.

The lowbush blueberry growth habit differs from the other species. Rather than appearing like a typical shrub with multiple stems growing from a crown, these plants produce aboveground, upright stems from underground stems known as rhizomes. Individual plants are not recognizable as distinct from other plants whose rhizomes grow nearby, but lowbush blueberries are still considered to be shrubs.

In the southern United States, low-chill blueberry species are utilized. Rabbiteye blueberry (*V. ashei*) is the species most commonly grown commercially. Most cultivars were developed in the latter half of the twentieth century from wild genotypes native to Georgia and northern Florida. The rabbiteye name relates to the appearance of the scar on the fruit where the flower was attached. Rabbiteye plants may grow naturally as high as 20 feet, but are usually kept to half that height or less in cultivated production. Rabbiteye stems may produce marketable crops as long as seven years. A number of other varieties and hybrids also exist for different locations and uses.

Most blueberry plants are deciduous and fall colors are typically a striking red, so people may plant the shrubs in their home landscape for ornamental value. All blueberries prefer a pH range 4.0 to 5.5, which is very different from most other crop plants that require a much higher pH. Blueberry roots tend to be very shallow. This makes it relatively easy to transplant established bushes to new locations, but they require a good amount of water to be successful.

Blueberry flowers are unusual, because the open end of the bell-shaped petal cluster points downward. The flowers are white and cream-colored, sometimes tinged with pink. Wild bees and bumblebees are the major pollinating insects, rather than honeybees, which are more common for many fruits.

Although the northern highbush blueberries will self-pollinate, having at least one other genetically different cultivar for cross-pollination increases yield for some cultivars. The southern highbush blueberries are somewhat self-pollinating, but produce more fruit when pollinizers are used. Rabbiteye and lowbush blueberries require pollinizers, due to their high degree of self-incompatibility. Blueberry fruits change color from green to pink to blue as they ripen. The fruits are actually false berries, rather than true berries. A true berry consists only of the

Figure 14.1. Blueberries are native to North America and are found primarily in the states of Maine and Michigan but also widely across norther Europe and Eastern Canada. Highbush blueberries are primarily grown as large, separate plants while lowbush plants are frequently found as widespread fields of closely clumped plants.

ovary and associated parts, like the seed. The blueberry fruit has vestiges of other flower parts in addition to the ovary.

Blueberries can stay attached to the plants after they ripen for longer periods than many other fruits, from several days to weeks. Any additional time after maturity increases the sugar content of the fruit. This benefit, however, has to be weighed against the greater chance the blueberries will be consumed by birds and other animals. For birds, blueberries are typically protected using netting or some type of scare device that makes noise or an imitation object that resembles a predator. For large animals, fencing may be the best option.

Harvesting occurs over a period of weeks, with multiple collections during that time. Fresh or frozen whole berries have been harvested by hand. Machine-harvesting is the primary method used for blueberries ultimately processed into baked goods, jellies, or juices. Due to their much lower height, lowbush blueberries are usually harvested using hand rakes, although some mechanical harvesters are available.

The United States leads the world in blueberry production, with Canada a distant second. Together, these two countries dwarf all other countries in blueberry production combined. In the United States, Maine is the major producer of lowbush, or wild blueberries. Michigan is the leading producer of cultivated blueberries, but other states are increasing production approaching that of Michigan.

Blueberries receive high marks for their nutritional value, as well as having relatively few calories per serving. In addition to having small amounts of vitamins and other minerals, blueberries contain antioxidants, which may benefit health in cancer prevention, slowing aging-related degeneration, and reducing infections.

b. cranberries

Like its blueberry relatives, this native plant with bright red to dark red fruit is a member of the Ericaceae family. Few cultivars are used in commercial production. Most are hybrids made from crosses of native plants. Cranberries grow on short, evergreen creeping shrubs, similar to the lowbush blueberries. They spread by rhizomes from which the upright stems grow. The stems grow slowly, eventually sagging downward at their bases from their weight. As a result plants extend only up to eight inches above the soil. Leaves typically remain on the plant for two years.

Figure 14.2. During most of the year, cranberry bogs are dry. It is usually only at harvest time that the bogs are flooded to facilitate cranberry harvest. Most cranberries are grown in Wisconsin, Massachusetts and New Jersey.

Due to its flowers' resemblance to a sandhill crane's head and neck, "craneberry" was the name given this fruit by early settlers to the United States. The open, pink-tinted petals point downward, similar to blueberry flowers. Among temperate fruit crops, cranberries are one of the last species in the season to flower. Cranberry fruits are classified as false berries for the same reason that blueberries are false berries. When cranberries first develop, they are green, changing to white before becoming dark red, to almost black when fully mature.

Cranberry production is similar in some ways to lowbush blueberry production, but also involves features unique to this crop. Acidic bogs or marshes in the northeastern United States and southern Canada are the native habitat for cranberries. Early production involved management of native stands. Most natural sites have been modified, while new beds are completely manmade. Artificial fields enable fruit production in western areas of the United States beyond Wisconsin.

One modification to natural stand management was the addition of supplemental water sources for the bogs. Three or more acres of supporting land per acre of cranberries contain water used for irrigation, harvesting, and frost and winter protection. In the past, irrigation was done mostly by flooding the bogs. The water was then drained from the field, because cranberry plants do not grow continuously underwater like some species of rice do. Sprinkler systems are more commonly used in current production systems. In addition to providing irrigation, sprinklers are used for frost protection when necessary. When water changes from liquid to solid, heat is released, which is enough to protect plants as long as the frost is not too severe.

At harvest time, fields are either flooded for a wet harvest or harvested "dry" with much less water on the field. Fruit collected using wet harvesting is processed immediately or frozen for later processing. Cranberries sold fresh must be dry-harvested so the fruit will not rot.

The machines used for wet harvests are designed to beat the berries off the plants. Fields are covered with 12 to 18 inches of water. The berries float to the water surface where booms are used to move the berries to the edge of the field for loading into trucks or crates. If crates are used, helicopters carry them to their destination for processing.

Dry-harvested cranberries are removed from the plants by machines with comb-like tines that do not damage the fruit. The freshest fruit has air pockets inside that allow the fruit to bounce. This phenomenon was first discovered by John "Pegleg" Webb, whose wooden leg made it difficult to carry the fruit down stairs. When he dropped the cranberries, he observed that the freshest berries bounced to the bottom of the stairs, while damaged fruit stayed on upper steps. A series of boards are still used today to separate the fruit by freshest quality. The fruits that bounce over all the boards are the ones packaged whole for fresh sale.

Flooding during the coldest months of winter protects plants from desiccation and excessively cold temperatures. Once the land is covered with ice, one half to one inch of sand may be applied to the ice surface. As the ice melts in the spring, the sand settles around the plants, encouraging rooting close to the tops of the stems. The sand also aids in pest control and providing a firm surface for equipment used during production of the crop.

Wisconsin and Massachusetts dominate production of cranberries in the United States. Cape Cod in Massachusetts was one of the first areas of commercial production. Today, Wisconsin harvests account for more than half of cranberries sold, with production mainly in central and northern portions of the state.

Cranberry fruit is tart, even when mature. Most cranberries are processed, rather than sold fresh. Numerous products are made from or contain cranberries. Dried cranberries are marketed as craisins.

More recently, some cranberries are harvested while they are still white, a few weeks before they would have turned red. These fruits are less tart and used to produce milder juices. An added benefit to the white juice is the avoidance of stains if the juice is spilled onto fabrics. Cranberries have similar nutritional and health value as blueberries. Cranberry juice is also promoted as beneficial for urinary tracts, helping to prevent urinary tract infections!

c. grapes

Fruits we know as grapes include species that are classified into one of two subgenera. The Euvitis subgenus includes species considered "true grapes." European grapes, (*Vitis vinifera*) and Concord grapes (*V. labrusca*) belong to this subgenus. The Muscadinia subgenus contains muscadine grapes (*V. rotundifolia*).

The true grapes are most productive in somewhat mild climates like that found near the Mediterranean Sea. Muscadine grapes grow best in warmer areas of temperate climates found in the southeastern United States where they are native. Several grape cultivars are named for the regions where they were selected for production.

Distinct morphological differences separate true grapes from muscadine grapes. True grapes grow in larger clusters of berries. True grapes also have thinner skins, forked tendrils, rough bark, and diaphragms in the pith at the nodes. Muscadine grape tendrils are simple, their bark is smooth, and there are no diaphragms in the pith at the nodes. At harvest, true grapes are collected by clusters, while muscadine grapes are picked singly.

European grapes account for more than 90 percent of grape production worldwide. Most of these grapes are made into wine. Other grapes in this species are consumed fresh (table grapes) or dried as raisins. The greatest European grape production occurs in California although many of the lower 48 states also have some

Figure 14.3. In the United States, native grapes grow in woodlands and along forest borders but they usually only produce small amounts of poor quality fruit.

commercial vineyards. Concord grapes are sometimes called American Bunch or Fox grapes and are turned into juices, jellies, jams, and preserves. Most Concord grapes are grown in the northeastern United States.

Although strawberries, raspberries, blackberries, blueberries, and cranberries have the word "berry" in their names, they are not true berries, because we do not consume just the ovary and its associated parts. The grape, however, does fit the definition of a berry, despite not having "berry" as part of its name. Grapes tolerate soils ranging from low to high pH. Deep, well-drained, sandy soils are the best for growing wine grapes. Too much soil moisture in wine grapes is so undesirable that irrigation is prohibited in some growing regions if grapes will be used to make wine.

The determining factor for vineyard locations is cold hardiness. Concord grapes tolerate the lowest winter temperatures, while muscadine grapes need the highest minimum winter temperatures.

Grape plants are woody vines, which must be supported by wires or trellising to keep them off the ground for ease of harvest, as well as keeping the berries cleaner. The vines attach to supporting structures via their tendrils. Each winter, up to 90 percent of the past season's growth is removed to maintain yield and quality of the vineyard.

Many grape cultivars are grafted onto rootstocks resistant to a root-feeding, aphid-like insect in the *Phylloxera* genus. This insect, believed to have been introduced from the United States to Europe, was the cause of major disruption to the wine industry in Europe in the mid- to late-1800s. Damaged plants appeared blighted and will not produce any more fruit.

Fruit maturity is determined by a variety of indicators. Table grapes are harvested based mainly on appearance of the berries. Grapes to be made into wine need to attain a certain level of sugar, measured in degrees Brix. Other factors include pH, acidity, and sugar-to-acid ratio. Concord grapes are harvested when their sugar level is still very low.

The skin of grapes contains the pigment that imparts color to the fruits. Anthocyanins are responsible for grape colors ranging from blue to purple, dark purple, and red. For wines, the length of time the skins remain with the juice determines how much tint will exist. White wines are produced by separating the skins and juice at the start of winemaking.

A wine "developed" in Rhode Island resulted from the consequences of Hurricane Bob that passed directly over the state in 1991. With the hurricane nearing, grape harvest at one vineyard was completed as hastily as possible. The usual orderly sorting of grapes by cultivars was abandoned due to time limitations, with all the cultivars being mixed together in one container. After the hurricane exited the state, the vineyard operation was left with a mixture of grapes normally processed separately. Since there was no other option, the vineyard's winemaker, or enologist, processed the mixed batch of grape cultivars. The resulting wine tasted so good, the "Eye of the Storm" is now produced yearly.

Figure 14.4. Commercially grown wine and table grapes are produced on moderately sized farms located in special environments that allow for the best quality grapes.

d. nutty fruits (pecans, walnuts, pistachios, cashews)

Nuts fit the definition of a fruit because we eat an associated part of the ovary—the seeds. Examples of common nuts are pecans (*Carya illinoinensis*), English walnuts (*Juglans regia*), pistachios (*Pistacia vera*), and cashews (*Anacardium occidentale*). Each of these nuts has unique characteristics. Most of the nuts produced commercially can be divided into two separate families: The Juglandaceae and the Anacardiaceae.

Juglandaceae Family

Pecans and walnuts occupy different genera in the same family, commonly referred to as the walnut or hickory family (yes, hickory, walnut, and pecan are very closely related!). All sixty species in the Juglandaceae family produce edible nuts, although pecans and walnuts are the most popular. Other *Carya* species include water hickory (*C. aquatica*), bitternut hickory (*C. cordiformis*), shellbark hickory (*C. laciniosa*), and shagbark hickory (*C. ovata*). Additional *Juglans* species include Southern California walnut (*J. californica*), butternut or white walnut (*J. cinerea*), and black walnut (*J. nigra*).

The chemical compound juglone is produced by plants in this family. It is especially high in the black walnut. Juglone is found throughout the plant. This compound is harmful to many other plants by killing or stunting them. The phenomenon where one plant produces chemicals that harm other plants is known as allelopathy. Juglone has been used as an herbicide and for its color value. Clothes, fabrics, and hair have been dyed with it. Some juglone has even been used in food coloring and inks.

Figure 14.5. Tree flowers (in this case, pecans) and grass flowers do not generally require bees for pollination. Instead, these plants utilize the wind to move pollen from flower to flower and look very different from what we normally consider a "flower".

Pecans

Georgia may be known as the Peach State, but it is also the number one producer of pecans. Texas' state tree is the pecan, although Texas is only the number three producer of pecans after Georgia and New Mexico.

Oklahoma and Arizona also produce significant crops of pecans. Some pecans are harvested from orchards using improved cultivars. Other pecans may be collected from wild stands of trees. The United States' harvest makes up more than 75 percent of the world's total of commercially grown pecans.

Pecan trees are native to the United States. The word "peccan" means "hard-shelled nut" in the Algonquin language. Trees grow best where soils have more sand than clay and are somewhat acidic, approximately pH 6.0. Roots grow very deeply, so areas with high water tables are undesirable. The southern United States' climate is best for growth, with many natural stands still existing. Some organic pecan production is practiced. Pecan trees are the tallest in the Juglandaceae family, reaching as high as 140 feet. The canopy may stretch as much as 80 feet in diameter. These dimensions make the pecan tree useful for shade as well as its wood, and pecan wood is often substituted for hickory or simply mixed together at the lumber yard.

Figure 14.6. Almost all of the pecan production in the world is located in the Southeast United states, where the tree is native. Pecans are very closely related to hickory, so much so that their lumber is often mixed together and indistinguishable.

Several pecan cultivars are available, with most established as grafted plants and some started from seed. Maturity is reached two to three years earlier in grafted trees compared to trees grown from seed. Seven years is typical for maturity in grafted pecan trees. These trees may remain productive for decades, which is much longer than many fruit trees. Due to their slow growth, growers find it economical to plant orchards at two or more times the final desired density. As trees increase in size, some trees are removed to avoid crowding.

Pecan flowers are monoecious. Male flowers are known as catkins. Both female and male flowers are borne on spike inflorescences. The flowers of both sexes are also incomplete because they lack petals and sepals. Pecans are dichogamous. This means the male and female flowers on the same plant are not active at the same time—ensuring cross-pollination. More than one cultivar is usually used for adequate pollination. Pollination can occur between trees miles apart, since wind is the method of pollen movement among plants.

The seeds and their shells are surrounded by a green, fleshy shuck. When mature, the shuck splits along four evenly spaced sutures. The pecan shell is smooth, oblong, and brown, with some cultivars having black markings. The seeds, or nuts, we eat are brown and their cotyledons have more uneven surfaces than cotyledons of many foods whose seeds we consume.

Mechanical trunk or limb shakers remove the nuts from the trees. A windrowing machine forms the scattered nuts into rows. The rows are then collected by sweeping machines. Blowers used during the latter two processes separate debris and leaves from the nuts. Some nuts are collected from tarps placed under the trees, rather than the use of windrowers and sweepers.

Figure 14.7. Most people have never seen a walnut on the tree. When they are first produced, walnuts are surrounded by a green fleshy, inedible husk. The walnuts fall off the trees or are harvested and the husks are removed revealing a nut. The husks do, however, produce a tenacious stain that has been used as a dye.

During processing, the shucks are removed from the shells. Nuts must be dried to 15 percent or less moisture before storage. Most nuts are shelled prior to sale and are used in baked products or candies. Although high in fat, the fat in pecans is monounsaturated. Additional compounds in the nuts help lower overall and LDL cholesterol and have antioxidant activity, all considered beneficial to our health. Even with the beneficial aspects, pecans are high in calories, so should be consumed in limited amounts.

English Walnuts

The English walnut is also known as the Persian walnut. This species produces the largest, most easily cracked nut of all the Juglandaceae species. Walnuts are native to Eastern Europe and Asia. China produces the most walnuts. In the United States, most walnuts are grown commercially in California. Dry, hot summers and mild winters, similar to Mediterranean climates, are optimal growing conditions. Soils should be deep, with little clay, and pH between 6 and 8.

Although not as massive as pecans, walnut trees may reach 100 feet or more. In orchards, they are maintained at one half or less their natural height. Most growers use grafted cultivars. Initial plant densities may be twice the final density to maximize profits during early production years. As plants increase in size, trees are removed to prevent crowding.

Like pecans, walnuts are monoecious with dichogamy. Male flowers are also called catkins. Walnut shucks are green and fleshy like pecans, but split unevenly, unlike pecans with their uniform sutures. The shells are brown and lumpy. The light brown seeds, or nuts, have more uneven cotyledons than pecans have.

Harvesting walnuts is done similarly to pecan harvesting. Processing is also similar, although walnuts are bleached because consumers prefer tan shells compared to the natural darker brown shells.

Most walnuts are sold shelled and used in baked goods and other foods. Walnuts are high in fat and calories, but limited amounts provide several nutritional benefits. The nuts are good sources of many essential nutrients, especially copper, manganese, and vitamin E. The monounsaturated fats may help reduce overall and LDL cholesterol. Some phenolic compounds may also have antioxidant properties.

Both walnut and pecan trees are hardwood species valued for making products like flooring, paneling, and furniture. Black walnut is preferred and while it ranges in color, it is typically a dark brown used frequently in rifle stocks.

Anacardiaceae Family

The Anacardiaceae family includes the edible nuts cashew and pistachio, as well as poison ivy (*Toxicodendron radicans*), poison oak (*Toxicodendron toxicarium*), and poison sumac (*Toxicodendron vernix*). What do these nuts and poisonous plants have in common that would place them in the same family? This taxonomic group, sometimes referred to as the cashew or sumac family, includes plants that have resin ducts in their bark, foliage, and/or fruits. Milky or clear exudates also may be produced. Most of the Anacardiaceae fruits are classified as drupes (stone fruits). Although marketed as nuts, cashews and pistachios are technically drupes.

Figure 14.8. The Chinese name for pistachio is "happy nut" and pistachios certainly are experiencing a resurgence in popularity. The two top producers of pistachio? Ironically, the happy nut is found predominantly in Iran and the United States, not the happiest of friends.

Cashews

The cashew is native to tropical portions of Central and South America. It is most abundant in Brazil. Other tropical regions in Asia and Africa now also commercially produce this species in even greater quantities than its native Brazil. Cashews are the leading tree-nut crop in the world. All cashews consumed in the United States are imported, since no commercial production occurs there.

Growing conditions require slightly acidic soil, preferably well drained. Cashews have no cold tolerance, which limits their production to tropical regions. High humidity and rainfall are detrimental to flowering and fruit development. The cashew is one of the few fruit crops usually grown from seed, rather than propagated vegetatively. Limited cultivars have been developed. The tree may grow to 40 feet, but is usually kept to less than half that height in commercial production. Cashew trees are capable of producing long, drooping branches, resulting in wide canopies.

Male and perfect flowers grow on the same inflorescence. The small flowers are red and may have stripes along their length. The petals curve backwards, exposing the stigmas and anthers for pollination. Flowering occurs over several weeks, which causes fruit maturation to occur over several weeks. Harvesting occurs after the fruit has fallen from the tree.

The nut and the peduncle of the cashew are edible. The nut is part of the true fruit. It is encased in a double-layered shell that resembles a boxing glove or kidney. Within this shell, an irritating oil known as urushiol is the same substance found in poison ivy, poison oak, and poison sumac. The urushiol must be removed before the shells are opened to avoid contaminating the nuts. This can be done by roasting.

The peduncle, which connects the fruit to the tree, is also called the "cashew apple," even though it is not a fruit like the domestic apple. Since the cashew apple is not derived from ovary tissue, it would be classified as a false fruit. The peduncle color progresses from yellow to red, when it looks somewhat like a real apple. Cashew apples detach from the tree along with the nut when the nut is ripe. Since cashew apples are highly perishable, they must be collected as soon as possible to avoid rotting. They are consumed fresh or processed into drinks, vinegars, or foods.

In addition to the edible parts of the plants, humans use the wood for building many products, including ships. The caustic oil, also referred to as CNSL (Cashew Nut Shell Liquid), is processed for industrial uses. Sap is used as an insecticide, as well as a constituent of some varnishes.

Pistachios

The second nut in the Anacardiaceae family, pistachios are native to Asia, where the Chinese translation for the name "pistachio" is "happy nut." Pistachios need long, hot summers for fruit production, but also have a long winter chilling requirement of between 600 and 1,500 hours to ensure proper flowering the following year. This species is drought- and salt-tolerant, but disease-prone if rain or high humidity occurs for extended periods during the growing season.

Commercial production in the United States did not start until the 1970's. Some climates in the southwestern United States provide the best growing conditions. California is the primary producer, with some production also in Arizona, New Mexico, and western Texas.

The pistachio tree may have one to several trunks, sometimes seeming more like a bush than a tree. Like cashews, pistachio trees may have broad, drooping canopies. Their width may be the same as the plant is tall, up to 30 feet. Commercially, pruning is used to facilitate the use of mechanical harvesters that shake the nuts from the trees. Pistachio trees are also used in ornamental plantings because of their growth habit and foliage. The compound leaves are large and appear somewhat gray. The longevity of the trees exceeds most other fruit trees. Pistachio trees may produce fruit for centuries under the best growing conditions.

The pistachio is a dioecious species. In the commercial industry, one cultivar, "Kerman" is used primarily as the female cultivar. "Peters" is the male cultivar usually utilized. Despite abundant flower production, fruit set may be only ten percent. Wind must move the pollen from the male trees to the female trees for pollination, which may account for some of the reduced fruit set, if the environmental conditions are not favorable.

While most nuts are harvested with their shells intact, pistachios are harvested with their shells partially split. At maturity, the skin of the nut changes from translucent to opaque. The hull around the shell changes from greenish to more reddish and becomes loosened from the shell. Shells that fail to split must be separated during processing so they can be mechanically opened. Nut colors range from yellow to green, with green being most preferred.

Delays in harvesting and initial processing may result in shells becoming stained by the hulls or infestation by navel orangeworms. Processing plants are located adjacent to or near pistachio orchards to facilitate hulling and drying the nuts within 24 hours.

Yields may be reduced by the presence of "blanks," or parthenocarpy, where fruits are produced, but no nuts develop inside. The blanks are mechanically removed during commercial processing. For homeowners, the easiest way to separate blanks from the shells with nuts is to place the nuts in water. Blanks will float to the top. The nuts should be dried as soon as possible if this method is used.

Trees may bear more fruits in alternate years, resulting in little to no crop every other year. During the heavy crop year, flower buds for the next year develop partially, but then abscise from the tree.

15

a. cucurbits

in chapter 13, we learned the definition of a fruit and a vegetable. We also learned that the definitions and differences between the two are both scientific and culinary. That is, scientists can call something a fruit, but because of its use, nutrition, and flavor profile, a chef will consider that very same item a vegetable. Hence, while a tomato is scientifically and botanically a fruit, tomatoes are treated like vegetables in cooking and in the grocery store!

One of the largest groups of fruits that are usually considered vegetables is the cucurbits. The cucurbit group is a very diverse group of plants that all belong to the family called the Cucurbitaceae. In total, the Cucurbitaceae has almost one thousand known plant species. Obviously, this family is very large and has many subdivisions, but just about all of the important crop plants that belong to this family are members of only a few genera. Most of these crop plants grow as annual vines, often producing tendrils, and they all have imperfect monoecious flowers; that is, each plant produces many separate male and female flowers. In most cases, plants will produce many more male flowers than female flowers. Because the female flowers will ultimately become the fruit and demand a significant amount of the plant's resources, the number of female fruit is limited. However, cucurbits will produce many male flowers because they require very little of the plant's resources and male flowers produce pollen that is necessary to pollinate female flowers. Bees are critical to this process. Without bees, few of the female flowers would be pollinated and fruit would either not be produced or be very small and inedible. Most cucurbits produce small yellow or white flowers and are highly attractive to bees scouring the landscape for nectar.

Many of the members of the cucurbit family produce melons (also called pepos). A melon or a pepo is a large fruit with a hard outer rind and soft, fleshy interior containing many seeds in the center of the plant. The most commonly encountered cucurbits in the United States belong to only a few genera (*Cucurbita*, *Cucumis*, and *Citrullus*), although many more exist that are either not typically utilized in food production or are considered exotic. Pumpkins, squash, and gourds belong to the genus *Cucurbita*, most commonly the species *Cucurbita pepo* (see the "pepo" in the name—now you know what it means!). Things get complicated at this point, however, because the species *Cucurbita pepo* has been bred into many different forms. An acorn squash is *Cucurbita pepo*. However, so is a zucchini. And so are many traditional Halloween pumpkins! Even though these things all look very different, they are the same species simply bred over long periods of time to have specific colors, textures, flavors, and shapes. An additional complication is the fact that not all pumpkins belong to *Cucurbita pepo*. Some pumpkins belong to *Cucurbita maxima*. And some pumpkins belong to other *Cucurbita* species. And some *Cucurbita maxima* varieties are not pumpkins but squash. This all begs the question, what is the difference between a pumpkin and a squash? The answer is, very little. Differences in color, shape, taste, and growing season do exist,

we eat these fruits but we call them vegetables

but these differences aside, these two crops are really just opposite sides of the same coin.

The *Cucurbita* genus originated in Central and South America and was domesticated by native peoples over thousands of years. While squash and pumpkins are frequently considered seasonally consumed crops, they were a staple of the Native American diet throughout Mexico and Central America. Some data suggest that these crops were in cultivation almost 7,000 years ago and have been moved and cultivated in portions of the eastern United States before its European settlement.

Figure 15.1. Pumpkins, watermelons and canteloupes are all in the cucurbit family but belong to different genera (the watermelon depicted is an heirloom variety called 'Moon and Stars' developed in the 1920's).

The genus *Cucumis* also contains many different cucurbits, with cucumber (*Cucumis sativus*) and muskmelon/cantaloupe (*Cucumis melo*) being the most common two crops in cultivation in the United States. While squash and pumpkin are almost always considered culinary vegetables and share most of the same essential traits, cucumbers and muskmelons are different. Most apparent is their shape and texture. Cucumbers are long and cylindrical and have a very thin skin. In fact, the skin of the cucumber is easily edible. Muskmelons/cantaloupes, however, are oval or round and have a thick rind that is not particularly edible at all. In addition to these differences, cucumbers and melons taste very differently from one another and are treated very differently in cuisine. Melons are considered fruit for their sweet and sugary taste. Cucumbers are considered vegetables for their mild and sometimes bitter taste (although some cucumbers can be a bit sweet). Despite these categorizations, they are both still botanically fruit.

Cucumbers originated in India and have been cultivated for at least the past 3,000 years or more. They traveled throughout the Middle East into Europe and were cultivated by the Greeks and Romans. Muskmelons and their relatives originated farther north, in areas of Africa to Iran and have been cultivated for at least 5,000 years, sharing a similar migration path to that of the cucumber.

The last major genus within the cucurbits is *Citrullus*. Although very similar to cantaloupes and muskmelons, watermelons are different, as their scientific name would suggest: *Citrullus lanatus*. Similar to muskmelons, watermelons are very sweet-tasting but are usually larger than muskmelons and have a smooth green rind (although this can be variable). In addition, the seeds of the watermelon are not contained in a central chamber but are dispersed throughout the fleshy inside of the pepo. Although the fleshy interior is used as a

Figure 15.2. Cucumbers come in just as many varieties as other fruits and vegetables. Here, pickling cucumbers, which as the name implies make excellent pickles, have been recently harvested. Pickling cucumbers also taste delicious fresh but their size and shape is perfect for a jar!

fruit, the watermelon rind is sometimes prepared as a vegetable. Watermelons were originally cultivated in southern Africa and have probably been domesticated for just as long as the other cucurbit species.

Most of the cucurbits evolved in tropical climates. Consequently, they often grow best in warmer climates with longer growing seasons. This is especially true of melons and watermelons that are difficult to grow in cooler climates. However, in warmer years a moderately sized fruit can be produced in some less-favorable environments. Pumpkins, squash, and cucumbers grow much better in cooler climates and will usually reach maturity before the frost. All of these species are very susceptible to frost, as would be expected of tropical fruit. Many of the cucurbits are grown in greenhouses, particularly those varieties that form bushes, as opposed to forming long clinging vines, which can take up a considerable amount of space. Cucurbits are susceptible to many different diseases and insect pests. Because each plant will usually produce only a few fruit, a severe disease can be catastrophic to the plant and to the people growing them.

Although the seeds of cucumbers are edible and the seeds of cantaloupes, squash, and pumpkins are easily enough removed en masse, watermelon seeds can be problematic. Because watermelon seeds are hard and inedible (you could eat them but they'll just pass right through your body) and produced within the watermelon flesh that is consumed, they can make eating watermelons messy. Fortunately, seedless watermelons exist that produce few seeds, most of which are white, soft, and edible pips. The reason seedless watermelons are seedless is that the plants are triploid (3N). In the process of meiosis, reproductive cells (which would be combined to form seeds) split their DNA up evenly so all cells will get the same number of chromosomes. When the reproductive cells of a pollen and ovary combine sexually, their even copies of chromosomes will recombine to form a complete whole. Most plants are diploid (2N). They have two copies of genetic material that can be split into the reproductive cell, which then become haploid (1N). When haploid (1N) cells combine, the amount of DNA will return to the original diploid (2N) amount present in the parent and the plant can continue its life cycle. However, a triploid (3N) plant has an odd number of copies of DNA—three copies. When it tries to split those three copies evenly for reproductive pollen and ovaries, it ends up with a mess. The result is that any seeds that are

Figure 15.3. Solanaceous plants all have very similar flowers but can be distinguished by their different colors: tomatoes are yellow, potatoes are pink and wild nightshades span a diverse range of colors from dark purple to white.

produced never mature because they have incomplete copies of their genetic blueprint and fail. Hence, any fruit produced on a triploid plant will be sterile and will not produce seeds. Some species and varieties of plants are tetraploid (4N) or hexaploid (6N) and triploid plants can be produced by crossing tetraploids (4N) and diploids (2N), as in the case with watermelon.

b. tomatoes

Tomatoes are a one of a number of very important and common crops belonging to the family Solanaceae. This group of plants is very large and is often called the nightshade family. As most people know, nightshade (or belladonna) is a poisonous plant and most of the plants in the Solanaceae produce some type of poisonous alkaloid or other potent molecule. Members of this family include peppers, petunias, tobacco, jimson weed, mandrake (which many people first became acquainted with through the Harry Potter series of books), and the nightshade. While petunias and tobacco are unlikely to be found outside of commercial agricultural production, jimson weed and nightshade, which both produce a highly toxic and usually fatal suite of compounds, can be found regularly throughout the countryside.

Tomato also produces at least two slightly toxic alkaloids: tomatine and small levels of solanine. Luckily, these products are usually present at very low levels and only in the green, inedible parts of the plant. Not only do these two alkaloids prevent the plants from being consumed by herbivores, but they are also mildly toxic to pathogens that cause plant disease. In addition to alkaloids, tomato fruit also contain high levels of many different vitamins and nutrients and the compound lycopene, which is considered a very useful antioxidant. The term "lycopene" comes from the old scientific name for tomato, *Lycopersicon esculentum*. Recent scientific work examining the DNA of tomato demonstrated that it really was a *Solanum* and thus a member of the potato genus. The current accepted scientific name for tomato is *Solanum lycopersicum*.

For nutritional and culinary uses, the tomato is a vegetable. But like the watermelon, it is a sweet-tasting fruit. In fact, the Supreme Court ruled in the 1890's that the tomato should be considered a vegetable for practical purposes of taxation. Tomatoes come in a vast array of shapes and sizes and even come in different colors ranging from ripe-when-green to purple, black, yellow, and white. Tomatoes are a New World crop, originating in Peru and the surrounding areas. They were domesticated and flourished in Central America and Mexico. Although tomatoes are traditionally associated with Italian cuisine, it was not until the Spanish arrived in Mexico in the 1500's that the tomato was brought back to the Mediterranean and became a staple of Southern European dishes.

Figure 15.4. Tomatoes come in a wide variety of shapes and colors, some are even purple!

Tomatoes are different from the cucurbits and many other crops in that they do not require bees or an outside source of pollination. While they do require warm climates for a successful crop, they are self-fertile. That is, each flower is perfect and contains both male and female parts that are compatible with each other. Left to their own devices, tomato plants produce fruit. And in many tomato varieties, the seeds produced from tomato plants are almost genetically identical to the parent plants because all of the DNA (from both pollen and ovary) is from the same parent. In most animal populations this would lead to inbreeding and other problems, but tomato plants do just fine with self-fertilization. Although they do not require bees, bees can increase fruit production by landing on the flowers and ensuring adequate pollen distribution.

Tomatoes are a unique crop for many reasons, but one of their peculiarities is their "weediness." While tomatoes are not particularly cold-hardy, tomato seeds can survive cold winters quite well, often germinating in the spring where old tomatoes lie decomposing in the soil or in compost piles. Tomato plants grow very quickly in warm weather and are either determinate or indeterminate. Determinate plants grow to a specific height then stop. Indeterminate plants continue to grow indefinitely and some varieties can reach 10, 50, or 100 feet if allowed to grow that long. Tomato plants do not like the cold; planting them too early in cool temperatures will often result in permanently stunted plants, but tomatoes transplant easily and can be abused badly and still recover well. Finally, tomato plants will produce adventitious roots all along their stem if grown sideways or buried deeply, growing "like a weed" even in poor soils.

c. eggplant

Eggplant is another solanaceous plant, even more similar to the nightshade than the tomato is. In fact, the eggplant is a type of nightshade, domesticated from wild Asian nightshade. The process of domestication was discussed in an earlier chapter but did not previously discuss poisonous plants. Obviously, no civilization cultivates poisonous plants for consumption, but in any population of poisonous plants, some individuals will be less poisonous or not poisonous at all through random mutation. Ancient populations inevitably identified these populations of plants and began rudimentary breeding programs. Small fruited plants that contained no poison were continually bred, looking for larger fruited plants, the ultimate result being the crops we grow today. Although this process is not one that we can directly observe because it happened thousands of years ago, it is the process we believe led to the domestication of poisonous eggplant progenitors and other plants like poisonous almond progenitors.

Figure 15.5. Aubergine is another name for eggplant, a fruit which is used in a number of different dishes but surprisingly, can cause allergic reactions in a small number of people.

So while the black nightshade (*Solanum americanum* and *Solanum niger*) produces extremely poisonous fruit, its domesticated relative the eggplant (*Solanum melongena*) possesses larger fruit of the same color, the same shape of leaves, and same general appearance, but is completely safe to eat. Despite their relatively safety, a small number of people do experience an allergic reaction to eggplant. This allergic reaction can usually be minimized if the eggplant is cooked or boiled, which is one reason people who are allergic may not realize it—most eggplant is consumed cooked. The process of cooking and/or boiling poisonous plants is a common method of detoxifying many poisonous plants, assuming that the toxins can be leached out of the plant or denatured by heat. Author's Public Service Announcement: **Do not eat wild plants. Do not eat wild mushrooms**. Hundreds of people die every year from eating wild materials that they assume are safe to consume but that happen to be extremely poisonous. It takes expert botanists, mycologists, and mushroom hunters decades of experience to identify edible plants. Don't be a guinea pig. Symptoms of wild plant/mushroom poisoning are varied but frequently irreversible (permanent liver damage requiring transplant) and frequently fatal.

Following the theme of this chapter, the eggplant is a fruit with culinary and nutritional vegetable characteristics. In fact, the fruit itself is bitter and rarely eaten raw, making it difficult for people to believe that it is actually a fruit. Most eggplant fruit are large, usually one to two feet in length, purple, and shaped like an elongated pear. As with most of the crops in this chapter, color and shape can be highly variable. Eggplant gets its name from some of the earliest varieties, which were beige or yellow in color and egg shaped. In

European countries the eggplant is called the aubergine. The eggplant is not typically a favorite in traditional American cuisine, but its use is common in a number of ethnic dishes.

Like most of the solanaceous plants, eggplant is a tropical crop and grows best in warm climates. It may be grown in temperate climates, but like other tropical crops, it is highly susceptible to frost and fruit may be undersized when grown in cooler regions. Similar to the tomato, the fruit are complete and can self-fertilize, but bees increase the level of fertilization and consequent fruit. Eggplant is an example of a crop that can be hand-pollinated (so are many of the larger cucurbits). In hand pollination, people ensure pollination. With a cucurbit, a person takes the parts of a mature male flower and applies them to a female flower, simulating the activity of pollinating bees. In a tomato or an eggplant, shaking the flowers vigorously is enough to ensure fertilization. This process is most effective in greenhouses or in gardens. Large-scale commercial field operations will often import beehives to ensure pollination. In fact, there is an entire industry of beekeepers (apiarists) who travel the country with flatbed trailers of honeybees, delivering hives to farmers to ensure pollination and the production of all kinds of different fruits! Without the efforts of these beekeepers, the cost of fresh fruits would skyrocket and many of our most common produce would simply be unavailable.

Figure 15.6. Green bell peppers are the most commonly encountered of the Capsicum peppers (not to be mistaken for the spice Piper pepper). Red and yellow bell peppers tend to be sweeter while the green peppers are spicier.

d. peppers

While there are many fruits called vegetables, the last one we'll cover in this chapter are peppers. Like tomato and eggplant, pepper is another solanaceous plant. The term "pepper" is actually used for a couple different types of plants. Bell peppers belong to the genus *Capsicum annuum*. Bell peppers are the large, fist-shaped sweet and spicy peppers you find in the grocery store and typically come in green, red, orange, and yellow. The green bell peppers tend to be spicier than the brightly colored bell peppers, which are often sweeter. As a result, some people call green peppers bell pepper and colored peppers sweet pepper. The red and orange peppers also tend to have higher vitamin, lycopene, and carotene content than the green peppers. From our quick survey of the Solanaceae, it is clear that green, yellow, orange, red, purple, and black are common solanaceous fruit colors and the same is true for the peppers.

The other pepper people encounter is the black pepper, which belongs to the species *Piper nigrum* and is a vine in the Piperaceae, a completely different family from the Solanaceae. Black pepper produces small hard black fruits

Figure 15.7. Some peppers are sweet, some are spicy. The Cubanelles are a sweetish pepper used in Puerto Rican cuisine, although the physically resemble the very hot and spicy jalapeño pepper.

known as peppercorns (another fruit that no one thinks of as a fruit) that are what you find in your pepper shaker. Peppercorns are ground up and produce a characteristic spicy flavor, very different from the flavor of bell peppers.

In addition to many colors of the bell pepper, there are many different varieties that have different sizes and shapes and are often given special names. It should be noted that the bell pepper is actually somewhat unusual as pepper plants go. Bell peppers are mildly sweet. This is not true of most *Capsicum* peppers, which tend to be spicy and hot. The "hotness" of a pepper comes from a compound called (not surprisingly) capsaicin. The chili pepper is a good example of a spicy or hot pepper. While most people could eat a bell pepper raw, eating a chili pepper raw would be painful. The capsicum produced in the chili pepper gives it a hot, spicy flavor that can be enjoyable in low to moderate levels, but painful at high levels. The reason bell peppers are not "hot" like chili peppers is that they do not contain capsaicin (interestingly, black pepper is spicy because it contains the alkaloid molecule piperine, which when consumed, may result in many different positive health effects). Some chili peppers are just varieties of *Capsicum annuum*. Other chili peppers are completely different *Capsicum* species.

Capsaicin has been used for many different purposes. While it gives many foods a spicy flavor, is it also used medically for a number of ailments including arthritis, psoriasis, fibromyalgia, and even some types of cancer. In addition, getting a face full of capsaicin is incredibly painful and is the main ingredient of pepper spray, which can be used in personal defense. It should be noted that pepper spray and mace are very different materials, but much of what is sold today as mace is actually either capsaicin or a mixture of capsaicin and other materials. The ability of capsaicin to act as an irritant is an evolutionary advantage of peppers. By producing capsaicin, a plant will deter its consumption by herbivores and not be gobbled up before it has a chance to seed. In contrast, birds do not respond to capsaicin and when they eat a capsaicin-containing fruit, they will actually spread its seeds! Finally, capsaicin does have some antifungal effects, protecting plants from disease. As you can see, capsaicin is a very important secondary metabolite.

Other *Capsicum annuum* varieties of pepper include the banana pepper, which is relatively mild-tasting, yellow/orange, and shaped reminiscently of a banana; the cayenne pepper, which is very hot when consumed, red, significantly smaller than a bell pepper, and shaped similar to a banana pepper; and finally the jalapeño pepper, which is small, green/red, and can be mild to hot in its spiciness. In addition to these subgroups within *Capsicum annuum*, there are even more species within the genus *Capsicum* (around 30 species) that produce a wide variety of unusual peppers—some of which are consumed and others that are used as ornamentals. And there are hundreds of different varieties of chili pepper, sweet pepper, bell pepper, and so forth. While it is easy to say that one type of pepper is hot and another is mild, every variety will have different characteristics so some chili peppers will be scorchingly hot and others will be extremely mild.

Like many of the solanaceous crops grown widely for consumption, *Capsicum* is a New World crop, native to South and Central America. While Native American populations have grown peppers, tomatoes, potatoes, and eggplants for thousands of years, these crops were unknown in European civilization until the 1500's when European explorers crossed the Atlantic. As they explored, colonized, and often pillaged indigenous populations, they discovered the native agricultural crops and brought them back to Europe and even Asia as global trade burgeoned. As mentioned earlier, people move plants very effectively and the introduction of solanaceous plants to Asia and Europe permanently affected the culture and cuisine of those societies, as well as spreading those plants to every corner of the globe.

16

a. the starchy potato

although many people in the United States believe that white potatoes (*Solanum tuberosum*) originated in Ireland or Idaho, they are actually native to South America and were a staple of the Peruvian Inca diets for thousands of years. Spanish explorers introduced potatoes to Europe and they crossed back and forth over the Atlantic Ocean in the 1500's and 1600's. At first, Europeans considered potatoes to be fit only for animal consumption, but gradually incorporated potatoes into their diets as a cheap but plentiful form of nutrition, especially among poorer families.

The Irish potato famine provides an excellent example of the danger of relying on one plant species as a major food source. Most of the potatoes grown in Europe were genetically similar. As a result of a devastating disease, late blight (*Phytophthora infestans*) to which all parts of the potato plants were susceptible, the crop was decimated across the country beginning in 1845. Many starving Irish people died or immigrated to the United States in ensuing years.

The potato is a member of the solanaceae family, as is the tomato and eggplant. Unlike the tomato and eggplant, where the fruit is consumed, we eat the stem tuber of potatoes. The swollen tuber typically forms at the end of a rhizome. In the commercial industry, the rhizomes are often called stolons, despite their technical growth as rhizomes. The starch stored in the potato supplies the food for the buds at the tuber's nodes. These buds, called "eyes," form the next year's plants. One potato can produce several plants, depending on the number of buds present.

Potato varieties vary in shape, size, and color. Larger, longer potatoes are preferred for baking and making French fries. Other varieties are almost as round as they are long. Potatoes harvested early in the growing season, when they are round and less than 2 inches in diameter, are marketed as "new" potatoes. Potato skin color ranges from brown to red, yellow, orange, blue, or purple. The skin texture may be fairly smooth to rough. Rough-skinned, brown potatoes are also known as russeted potatoes. Internal colors vary as much as skin colors. Some potatoes with golden flesh taste somewhat buttery by themselves.

Commercial growers plant their crop using "seed" potatoes, which are actually small whole potatoes or tubers that have been cut into several pieces by hand or mechanical cutters. Each piece has up to three buds and a large enough section of the starch-containing cells surrounding the bud(s) to provide food for the developing plant until it emerges from the soil and can make its own food from photosynthesis. This method of vegetative propagation assures genetically identical potato plants will produce tubers of similar size and eating quality. Although grown as an annual, in climates with milder winters, tubers may survive in the ground and produce plants the following spring.

Seed potatoes planted immediately after cutting tend to be infested less frequently by microbial pathogens like bacteria and fungi. If not planted immediately, the potato pieces are allowed to dry around the cut edges (suberized) and stored in a cool, humid environment. Seed pieces may be dusted with a fungicide to protect the pieces from soil-borne fungi. The potatoes are established in rows at densities based on cultivar, environmental conditions, and desired potato size. For example, potatoes used by the French fry industry are spaced far enough apart to grow long tubers.

Either before or after planting, ridges or hills are formed in the fields. The furrows formed by the mounding process promote adequate drainage and allow directed flood irrigation, where utilized. The soil mounding encourages increased tuber development. It also helps prevent exposure of tubers to light, where they turn green due to chlorophyll production. A toxic alkaloid, solanine is also produced when any part of the potato turns green. Solanine imparts a bitter taste, especially in the green area of the tuber. Cutting off green sections and cooking remove most of the solanine present in affected potatoes.

Figure 16.1. While potatoes are not grown for their fruit, they do occasionally produce poisonous green berries that are very similar to tomato fruit and even have very similar flowers, which are lavender instead of yellow.

Mechanized equipment is used throughout the production process. Potato planters drop seed pieces at desired intervals. Irrigation may be applied with sprinklers or by flooding furrows. Harvesting equipment lifts the mature tubers from the ground and conveys them to trucks for transport to processing centers. For stored potatoes, vines should be mature to ensure tougher skins are formed on the tubers.

Pest control is essential for maximum yields. Numerous insects attack potatoes directly, as well as transmitting damaging diseases. One of the most serious insect pests is the Colorado potato beetle. This insect has become resistant to many insecticides. In the home garden, hand removal or vacuuming may be sufficient control measures for the Colorado potato beetle. Many of the diseases that affect potatoes may be avoided by using resistant cultivars, disease-free seed, and crop rotation. Where used for weed control, cultivation should be shallow enough to avoid damaging the developing tubers.

Homeowners may purchase seed potatoes or make their own. Since potatoes sold in most stores have been treated with a chemical to prevent sprouting, those potatoes must be stored long enough for the effect of the treatment to diminish. And unfortunately, while store-bought potatoes can be used to start new

Figure 16.2. Most people are familiar with the ubiquitous brown or beige "white potato" but potatoes come in many different varieties, including this purple variety called 'Vitelotte'.

potatoes in the fields, they are often infected with some level of disease that can be transmitted to a homeowner's garden (plant diseases don't cross over to people so they can't hurt us if we eat them but they can spread to other plants and damage garden plants). If buds have begun to grow, the potatoes can be prepared for planting the same as done by commercial growers. Home garden production of potatoes often simulates commercial production. Gardeners also mound soil around plants to increase the depth of the growing area for the potato tubers. A garden fork pushed into the ground so the tines are below the tubers is usually used to lift the tubers to the surface. Undamaged potatoes may be stored in dark, cool, humid conditions.

Potatoes are not grown from actual seeds because the plants will not be genetically identical. The harvested crop will likely have potatoes with variable appearance and eating quality. Only plant breeders grow their plants from actual seed in order to use this variability in selecting parents for making crosses to produce new cultivars. One famous plant breeder, Luther Burbank, developed the Burbank potato in the mid-1800s, a cultivar widely grown to this day and favored for French-fry production.

"Meat and potatoes" describes the two foods around which many meals are designed. Potatoes contain fiber, as well as many essential nutrients and minerals. The versatility of this vegetable ranges from servings of baked or boiled whole tubers to scalloped dishes from sliced potatoes, hash browns from shredded potatoes, and French fries from chopped potatoes. Potatoes may be consumed fresh, frozen, canned, processed into products such as potato chips or dehydrated as potato flakes for later reconstitution as a number of dishes, including mashed potatoes and potato pancakes.

b. salad staples (celery, lettuce, etc.)

The phrase "garden salad" conjures up various images, depending on one's preferences. Vegetables whose leafy green portions serve as the main ingredients of garden salads are also simply called greens or salad greens. Leaves from several species may be used when mixing salads. Some leafy greens are also served cooked. Additional plant parts from leafy greens may be incorporated into salads or other food preparations. These parts may include seeds, flowers, stems, or roots. Some non-leafy vegetables are incorporated into garden salads to add color, flavor, and texture. Examples of leafy greens and non-leafy vegetables are included in this section.

Figure 16.3. Many different varieties of lettuce exist, including some that even have red leaves.

Lettuce

While lettuce (*Lactuca sativa*) has been the dominant green leafy vegetable used as the base in many salads, other non-lettuce leafy greens are increasingly utilized. The non-lettuce greens enhance the color, flavor, and texture of garden salads. Surprisingly, lettuce is a member of the sunflower family, or Asteraceae, and originated in the Mediterranean. Lettuce seeds look very similar to dandelion seeds, which also belong to the sunflower family and many wild (and barely edible) species of lettuce exist. Although not a commercial salad green, every part of the dandelion is edible and they often find their way into ethnic salads. Lettuce is grown as an annual, although it is a true biennial. Leaves vary from a pale green to deep purple, depending on variety. Three types of lettuce are based on their shape: head, Cos or Romaine, and loose-leaf.

Head lettuce grows as a group of leaves formed into a somewhat rounded shape. The most well-known head lettuce type is Iceberg lettuce. This lettuce is less nutritious than many other salad greens, but is favored for its crispness and longer storage life compared to other greens. Bibb and Boston are other types of head lettuce. They form loose heads because their leaves are less densely packed than leaves of Iceberg lettuce. These varieties are also known as butterhead varieties because their leaves are sweeter and tenderer than Iceberg lettuce. Cos lettuce leaves form elongated heads. Their large leaves are favored for sandwiches as well as salads. Cos lettuce is the main ingredient of Caesar salads. Loose-leaf lettuce plants produce several leaves that do not form heads. Varieties vary in color from light green to reddish-green with curly leaf margins. These attributes enhance the appeal of salads.

Figure 16.4. Lettuce is an early season crop that does best in cool conditions.

Lettuce grows better under cooler temperatures, typically between 60 and 65°F, and can be started in fields much earlier than other crops. While some lettuce, like Cos varieties, tolerate higher temperatures, other varieties may flower and produce seed with temperatures 5 to 15° F above optimum growing conditions. Lettuce plants contain latex as a healing and defense mechanism. Unfortunately, latex is bitter and lettuce plants that flower produce a lot of latex and are too bitter for sale. To attain the high water content of lettuce leaves, consistent, ample water availability is essential. Lettuce plants tolerate various soil types, but grow best on fertile, slightly acidic, well-drained soils containing high amounts of organic matter (particularly organic muck soils).

Seeds or transplants are used to establish head lettuce. Spacing depends on expected head size and equipment used for cultivation. Leaf lettuce is typically seeded in three or four close lines within each row. The greatest commercial production in the United States occurs in California.

Hand-harvesting is used for all lettuce types grown for sale. Multiple harvests may be necessary to obtain optimum head or leaf sizes. Tight head-lettuce types like Iceberg are packed directly into boxes after they are cut from the soil. Butterhead lettuce may be packed like Iceberg lettuce or placed into plastic sleeves before packing. Entire plants are harvested with leaf lettuce varieties. These plants are often encased in plastic sleeves prior to packing. Leafy greens, like lettuce and herbs, should be torn by hand, rather than cut to release the best flavors of these foods.

Spinach

Spinach (*Spinacia oleracea*) is dark green, leafy staple of garden salads belonging to the Amaranthaceae family and originating from central Asia. This leafy green has a more distinctive flavor than lettuces. The leaves have been described as tasting somewhat sharp or bitter with a slight saltiness. Darker leaves tend to contain higher levels of nutrients, including vitamin C, but growing conditions are similar to those of lettuce.

The relatively short growing cycle of four to seven weeks for spinach enables growers to harvest the spinach again after two or three periods of regrowth. Alternatively, other crops may be grown on the same land during the warmer weather less favorable for spinach. Double-cropping spinach fields with other vegetables maximizes profits. California produces the greatest amount of fresh spinach. Texas tops the list of states producing spinach for processing. Processed spinach is either canned or frozen and served cooked, rather than used in salads like fresh spinach.

Younger spinach leaves are harvested by hand or machine. Hand-harvested spinach is sold in bunches that are not always washed prior to sale. Machine-harvested spinach is sold bagged after washing and sorting. The youngest leaves are marketed as baby spinach, while the next older leaves are called teen spinach or just spinach. These are most ideal for salads, because they are tenderer than older leaves.

Other Leafy Greens

Arugula (*Eruca sativa; E. vesicaria* subsp. *sativa*) is known for its pungent aroma and peppery taste, similar to related species in the Brassicaceae family (the mustards) to which it belongs. Originating in the Mediterranean, this foot-tall annual grows best in cool temperate climates. In warmer zones, arugula leaves may taste bitter.

Most people grow arugula from seed, since it does not transplant well. While tolerant of full sun, partial shade helps the plant tolerate the hottest part of the summers. Arugula should be harvested by pulling leaves from the base of the plant, rather than cutting them. In salads, arugula is usually mixed with other greens because its flavor may be overwhelming if used alone. Flowers, young seed pods, and seeds also are edible.

Curly Endive and Escarole (*Cichorium endivia*) is another Asteraceaeous plant that imparts a strong, somewhat bitter taste to salads. They are typically used to accent leafy green salads, rather than serving as the base for the salad. The outer leaves of curly endive heads are greener and bitterer than the somewhat blanched inner leaves. Endive leaves are narrower and curlier than escarole leaves. Escarole tends to be less bitter than curly endive. Endive and escarole production is similar to lettuce.

Celery

Celery (*Apium graveolens*) adds extra crunch to salads and belongs to the same family as carrots, the Apiaceae. This biennial vegetable is grown as an annual. Several leaves attach to a compressed stem or crown. The stalks are actually petioles supporting the lacy leaf blades.

Celery is usually planted using transplants to assure uniform spacing and allow the maximum growing time for this slower-growing vegetable. Soils should be moist, yet well drained, with neutral pH. During production, celery plants, known as bunches, may be tied or otherwise secured to shield the outer stalks from becoming too green for consumer preferences. This process is known as blanching. Self-blanching cultivars do not need any shielding. Consumer tastes are changing with some favoring greener petioles. Novelty varieties have been developed that have red or pink petioles.

Celery plants are harvested by hand or machine before the petioles develop tough pith. The plants may be sold as entire bunches or with the outer stalks removed, leaving only the more tender inner petioles, sometimes marketed as the "hearts." Rapid cooling is essential for optimal storability at cold temperatures and close to 100 percent humidity for two to three months.

Stalks are chopped into small pieces for use in salads. Several processed food recipes also include chopped celery. This vegetable provides fiber and nutrients essential to human diets. Unfortunately, a small percentage of the population is highly allergic to celery and allergic reactions can be as severe as allergic reactions to peanut in susceptible individuals.

Carrots

For color, carrots (*Daucus carota*) are often included in garden salads but we eat their tap roots, not their leaves. Cultivated carrots as we know them are typically orange, but they were developed from their wild relative of the same genus and species, which had purple roots. Over time, plants with orange, yellow, or white roots were developed by selection and breeding. Although orange roots are preferred by most consumers, other cultivar colors, including purple, are now available.

The wild relative, native to Eastern Europe or Western Asia, has the same genus and species name as the cultivated carrot. Another name for the wild relative is Queen Anne's lace. Many of the wild carrot plants found today in the United States have light-colored roots, instead of the purple color of the original species. This may be the result of plants once selected for lighter root colors reverting to wild types with thinner roots, but lacking the purple pigment after escaping from domestic cultivation.

Figure 16.5. Carrots are orange because of the pigment in their roots: carotene.

Caution should be used if one decides to taste wild carrot roots. In fact, it's best not to eat any wild plants or mushrooms unless they have been vetted by an experienced plant hunter. In this case, proper identification is essential because a poisonous species with a similar inflorescence is spotted water hemlock (*Cicuta maculata*). This carrot lookalike has less-lacy leaves with dark splotches. All parts of wild hemlock are deadly.

Carrot cultivars vary in size and shape. Some cultivars produce roots with larger diameters nearest the soil surface, tapering to a point. Other cultivars' root diameters remain the same along the length of the root. Root lengths range from roughly four inches to more than seven inches. Biennial carrots are grown as annuals. They grow best in cool temperatures between 60 and 70°F. Excessively hot weather may cause the roots to develop poorly and taste bitter.

Carrots should be harvested while the roots are still tender and sweet. Older roots become fibrous, or woody, and may have a bitter flavor. Most commercially grown carrots are mechanically harvested by undercutting the roots and lifting them from the ground. At the processing plants, roots are cooled, washed, and sorted mechanically by length and diameter, before culling by hand for defects.

Carrots are marketed fresh or processed by canning or freezing. Fresh carrots may be stored up to nine months with proper cold temperatures and 100 percent humidity. When used in salads, raw carrots are typically shredded. This vegetable is notable for its high beta carotene, a precursor of vitamin A. Darker roots have higher nutrient levels than lighter roots.

Radishes

Another vegetable added to salads for contrasting color is the radish (*Raphanus sativa*). This vegetable is a member of the Brassicaceae family, a grouping most known for cabbage, described in the next section, and mustard, discussed in Chapter 17. Like the carrot, the tap root of radishes is the plant part consumed.

Some radishes have round swollen tap roots. These globe types grow rapidly and are usually harvested when they are approximately 1 to 1½ inches in diameter. Most of the root is red, with the distal end white. In salads, these radishes are usually sliced without peeling to retain the distinctive red skin color. Other radishes have long, white taproots similar to the roots of some carrots. White radishes are most similar in size and shape to the carrot varieties that have uniform diameters along their length.

The Japanese radish, or daikon, is one of the largest varieties of white radishes. It may be more than one foot long and greater than two inches in diameter. The daikon is most often used in soups, grated and served with sashimi (raw fish), or pickled, rather than added to garden salads.

Figure 16.6. Radishes come in a variety of colors including purple, red and white (a daikon).

Radishes are established from seed that may be treated with hot water or fungicide to reduce crop loss from seed-borne diseases. They may be grown in many types of soils, avoiding rocky or heavy soils that may impede proper root development. Increasing pH above 6.5 reduces the incidence of club root (*Plasmodiophora brassicae*), a disease that severely deforms roots. Disease pressure may also be reduced by using crop rotations that avoid planting Brassicaceae crops sooner than every four years in the same location.

Cultural practices optimize root development. Like carrot production, excessive nitrogen fertilization must be avoided in radish production, because it will promote shoot growth at the expense of the desired root growth. Early plantings of globe radishes may reach harvestable size fast enough to not need supplemental irrigation. Slower-growing radishes, like daikon, and crops planted later in the growing season, when rainfall may be scarcer, are likely to benefit from irrigation.

Despite being in a different genus than cabbage, radish has been successfully crossed with cabbage. The resulting plant was named a rabbage. This hybrid had leaves resembling radish leaves and roots like those on cabbage, leaving it undesirable for further development into an edible plant.

c. the same genus and species: cabbage, broccoli, brussels sprouts, and cauliflower

Cabbage, broccoli, Brussels sprouts, and cauliflower may look different from each other, but they are the same species, *Brassica oleracea*, in the Brassicaceae family. The origin of this species is not well known, although it was used throughout Europe in the Middle Ages. The Brassicaceae family is commonly referred to as the mustard family. Flowers in this family have four white or yellow petals, whose shape resembles a cross. This family was once named the Cruciferae family because *cruciferae* in Latin means "cross-shaped."

Plants in *B. oleracea* are sometimes referred to as the "cole crops." Some of these species are biennials grown as annuals. Other species are true annuals. Some *B. oleracea* types are both cold hardy and store well. Distinguishing features separate the cole crops into designated groups (Table 16.1).

Table 16.1. *Brassica oleracea* groups.

Group	Common name
Acephala	kale, collard
Botrytis	cauliflower, heading cabbage
Capitata	cabbage
Gemmifera	Brussels sprouts
Gongylodes	kohlrabi
Italica	sprouting broccoli

Due to their close genetic relationships, *B. oleracea* plants in different groups may be cross-bred. Broccoflower, for example, contains genes from broccoli and cauliflower. The curd, or inflorescence, most resembles cauliflower in shape and texture. The dark green broccoli flower bud color and the creamy white cauliflower florets produce a lime-green head in their hybrid. Broccoflower flavor has been described as milder than either of its parents. This hybrid is occasionally available in markets.

Cabbages

Cabbage is usually pale green to medium green, although red cultivars exist. Its leaves wrap around each other in dense layers, forming a round, pointed, or somewhat flattened head, depending on cultivar. The red color in some cabbages results from high levels of anthocyanin pigment.

Different types of cabbage are suited for planting in one or more of the four seasons of the year in temperate zones. Generally, cabbages grow best in cool weather between 60 and 68°F. They can tolerate some temperatures below freezing. Cabbage plants exposed to periods of colder weather, followed by excessive heating may flower prematurely (bolt) during their first year of vegetative growth. The long flower stems render the cabbages unsuitable for harvesting.

Figure 16.7. Cabbage is a member of the species *B. oleracea*, which also includes broccoli, Brussels sprouts, and cauliflower.

Due to their genetic similarity, the cole crops are susceptible to many of the same insect and disease pests. Crop rotation with non-cole crops for three or more years between cole crops is recommended to help reduce pest incidence, particularly soil-borne diseases in later crops. Heads are harvested when they reach their mature size. Cultivars vary in uniformity of time for heads to reach maturity. More uniformly growing cultivars may need only one or two harvest sessions to clear commercial fields. Less-uniform cultivars may require three or more passes through fields before all heads are removed.

Heads are cut by hand at the soil surface. The harvested produce is removed from the field by hand or conveyor for packing in boxes. Due to their high moisture content, harvested cabbages should be removed from the field and cooled as quickly as possible to avoid excess water loss from the leaves. At cold temperatures near freezing and high humidity, some cabbages may be stored five months or longer.

Cabbage is served raw, cooked, or fermented. Raw cabbage is usually served chopped in salads. Cooked cabbage is often used in soups or stews. Corned beef and (cooked) cabbage is a popular combination in Irish cuisine. Fermented cabbage is known as sauerkraut and prepared in a variety of ways in Korean kimchi. Chopped cabbage is anaerobically processed in salty brine. Cabbage contains high levels of beta carotene and vitamin C.

Broccoli

The plant part consumed from broccoli consists of a dense arrangement of immature flowers on stems of a half-inch or more. Both flowers and stems are edible. Inflorescence colors range from the most commonly grown dark green to lime-green and purple. The growing season for some cultivars is long. Most commercially grown broccoli is directly seeded into the fields. Homeowners may use seed, although transplants are usually available for earlier harvesting.

Broccoli plants require wide spacing of up to two and one half feet. Their growing conditions are similar to cabbage. At harvest, heads should have buds that are small and tightly closed, with no yellow petals showing. Broccoli heads have short storage lives of two weeks compared to months for cabbage heads. Some cultivars have side stems that will produce smaller, harvestable heads for several weeks after the terminal flower head is removed.

Brussels Sprouts

This variant of the *B. oleracea* plants was discovered as a sport (mutation) on a cabbage plant in the Brussels region of Belgium. Although biennial, Brussels sprouts are grown as annuals. Several miniature cabbage-like heads grow along a vertical stem. Taller stems may require staking for support. Like cabbage, leaves may be green or red. Early, middle, and late cultivars are available. Sprouts are harvested when roughly the size of a walnut. Production and storage conditions are similar to cabbage.

Cauliflower

Cauliflower may be annual or perennial. The immature flower heads, or curds, are eaten. Heads are usually creamy white, but also may be purple or yellow-green. Purple cultivar heads turn green after cooking. To promote the whitest heads, blanching may be done by tying leaves around the heads as they develop. Self-blanching cultivars do not require this additional labor.

Optimal growing conditions for cauliflower are similar to those used for cabbage production. Misting during hot, dry periods in the growing season may be necessary. Harvested cauliflower may be sold fresh or processed by canning or freezing. Fresh heads last up to five weeks when stored under cold temperatures and high humidity. Fresh cauliflower may be eaten raw or cooked. The florets are frequently served as crudités.

d. peas and beans—good for us and our gardens

The terms "peas" and "beans" are commonly applied to several genera and species in the Fabaceae family, formerly known as the Leguminaceae or legume family. This section includes only peas that belong to the *Pisum* genus and beans associated with the *Phaseolus* genus.

Plants in the Fabaceae family have a symbiotic association with a bacterium in the *Rhizobium* genus. The *Rhizobium* bacteria convert nitrogen from the atmosphere to a form that can be used by the plants. In return, the plants produce more food for themselves, as well as enough to support the bacteria. Plant material that remains in the field after crop harvesting releases nitrogen into the soil as it decomposes.

Fabaceae plants have irregular five-petaled flowers. Their mature fruit is a dehiscent pod. Legumes are a significant source of protein in human and animal diets. Other species in this family include alfalfa (*Medicago sativa*), clover (*Trifolium* spp.), soybeans (*Glycine max*), and peanuts (*Arachis hypogaea*).

In addition to pod and seed harvesting, legumes are also grown for use as a forage (animal food) and as green manure crops. No pods or seeds are harvested from plants grown for green manure utilization. Before seeds form, plants are tilled into the soil for decomposition and release of nutrients. These nutrients, especially nitrogen accumulated from *Rhizobium* fixation, become available for subsequent crops that do not have nitrogen-fixing capabilities. In this way and others, green manures improve the quality of the soil for future years.

Peas

Native to the Middle East, peas have been grown for more than 9,000 years. People consume edible pods, immature shelled peas, and dry shelled peas. Garden peas (*Pisum sativum* subsp. *hortense*) include types known as snow peas and snap peas. China peas (*Pisum sativum* var. *saccharatum*) and sugar snap peas (*Pisum sativum* var. *macrocarpon*) also are grown for pods and shelled peas. Sweetness of the pods and seeds varies among varieties.

Europe and northern Asia produce the most fresh pea pods. The United States leads the world in fresh shelled pea production. Within the United States, fresh peas are grown across

Figure 16.8. Sugar snap peas are grown widely in backyard gardens for their sweet taste, eaten in their pods. Peas climb with their tendrils so they require a substantial physical support to be successful.

the northern states. Much canning of fresh shelled peas occurs in the Midwestern United States, particularly Minnesota and Wisconsin. Russia and China produce the most dried shelled, or field, peas. In the United States, the Northern Great Plains region leads in dry shelled pea production. Most of these peas are exported.

Cool temperatures produce the best pea crops, whether pods or seeds are desired. Optimum temperatures range from 55 to 65°F. Plants may even tolerate some frost. High temperatures, especially during pod and seed development, adversely affect pea yield and quality.

Seeds may be inoculated with *Rhizobium* bacteria to increase nitrogen fixation. The *Rhizobium* strain is different for each legume crop. Additional nitrogen is usually supplied through fertilizers to further increase crop yields. Maximum height for peas ranges from two to five feet. Taller peas are generally staked if grown in home gardens and sometimes in commercial production. Modified leaves, in the form of tendrils, wrap around supports such as stakes or trellises.

Peas for edible pod harvest are grown directly on the ground, in raised beds, or on trellises. Trellised snow and sugar snap peas tend to have fewer fungal problems, are easier to harvest, and have cleaner pods than peas that are not trellised. Commercial plantings that use trellising require significant labor, so they are usually grown on smaller acreage plots.

Multiple crops of edible pod peas may be produced on the same land in a single growing season, especially where climates are coolest. An alternative would be double-cropping with another crop that tolerates higher temperatures less favorable to pea production. Crop rotation to avoid growing peas several years in a row on the same plot may help prevent buildup of diseases and insects that decrease yield and quality of the peas.

Pods are graded for uniformity, lack of fullness (no significant seed development), lack of black spots (indicating fungal incidence), and calyxes that are fresh and green. Sugar snap pea pods have some swelling when harvested, while the other types are fairly flat. Higher prices may be paid for smaller pods that tend to be sweeter. All peas noted above are grown for their edible pods. Snow pea pods are popular in cooked Chinese dishes. Pods of sugar snap peas are eaten both fresh and cooked.

Mechanization of seed establishment and harvesting greatly reduced labor required for shelled pea production. Most shelled pea production is not trellised. Peas are harvested when pods are plump. Rapid cooling of the peas is essential as soon as possible after harvesting to avoid overheating in the containers. Much of the shelled pea crop is grown under contract with processing companies. The goal is to match yield with the companies' production capacity.

Green peas are graded by size, with smaller peas garnering higher prices, because consumers like sweeter peas. Smaller seeds tend to contain relatively more sugar than larger, more mature peas whose sugar is increasingly converted to starch. Green peas are canned or frozen, if not consumed freshly picked.

Dry shelled peas have less than 16 percent moisture to prevent disease development during storage. They are marketed as either green or yellow varieties. These seeds are a good source of protein for both people and animals. Premium peas for human consumption are graded for seed size and shape, splitting potential, shriveled seed, dirtiness, and bleaching. Bleaching may result from excess rainfall or humidity, bright sunshine, and warm temperatures at maturity.

Dry edible peas may be stored many months, providing a food source beyond the growing season. Seeds are rehydrated in liquid before or during cooking. Pea soup is a popular way to prepare dry edible peas for a meal.

Beans

Many bean species and varieties of species are utilized for human or animal food. The *Phaseolus* species are warm-season plants native to Central and South America. Growth habits include pole or runner, half-runner, and bush types. Pole and runner bean cultivars have indeterminate growth, while half-runner bean cultivars have semi-determinate growth and bush bean cultivars have determinate growth. Pole bean plants may grow as long as ten feet. Bush bean plants typically range from one to two feet in height. Half-runner plants are intermediate in size to the other two types.

Phaseolus vulgaris is often called the common bean. Green snap or string beans are the second most popular garden vegetable after tomatoes. The "string" in its name refers to the tough string growing along one seam of the pod in older cultivars. Newer cultivars have been bred to reduce the presence of this structure. Green beans and wax, or yellow, beans are harvested as pods with immature seeds inside.

In addition to green and wax beans, this species includes several varieties consumed as dry shelled beans. Bean preference varies among cultures. In some areas, the color of the bean is as important as the flavor of the bean. Dry edible beans, also known as field beans, are favored for their storability as well as their flavor. Some of the many varieties are included in Table 16.2.

Lima beans are categorized by size. Smaller lima beans (*Phaseolus lunatus*, Sieva type) are often called baby lima beans. Larger lima beans (*Phaseolus lunatus* var. *macrocarpus* or *Phaseolus limensis*, Lima type) are also known as butter beans, except in the southern United States where the smaller lima beans are sometimes called butter beans, with the larger beans simply referred to as lima beans. Lima beans are more completely described in Table 16.2.

Runner beans belong to the species *Phaseolus coccineus*. They are more specifically known as scarlet runner beans because of their flower color. The attractive nature of this species makes it valuable as an ornamental as well as a vegetable. Pods, dry beans, and tubers are edible.

Phaseolus beans are self-pollinating, so single cultivars may be planted in one location. Commercially, beans are planted from seed. Homeowners typically plant beans from seed as well, but transplants may be available to shorten the time to harvest the first fruits.

Seeds of beans grown for harvesting pods or immature seeds are usually not inoculated with *Rhizobium*, resulting in minimal nitrogen fixation. Seeds grown for dry bean harvesting are more likely than fresh beans to be inoculated with the *Rhizobium* bacteria, since dry beans have longer growing seasons to allow more nitrogen fixation.

Snap, wax, and some lima bean pods are harvested with immature beans inside. Snap and wax bean pods usually show little if any swelling from the developing seeds within them. Some people prefer beans with enough seed development to make them appear lumpy.

In the home garden, snap and wax bean plants continue to flower and produce more pods as long as they are harvested to keep seeds from maturing. Snap beans require fewer growing days than lima beans harvested in their immature state, primarily because the snap beans are harvested when their immature seeds are smaller than immature lima bean seeds.

Fresh bean harvest time is determined by pod size. Most fresh beans are canned or frozen because they do not have long storage lives. Florida, Georgia, and California lead the United States in growing fresh snap bean produce. Snap bean production for processing occurs mostly in Wisconsin, Oregon, Michigan, and New York.

Snap and wax beans are eaten plain, pickled, and more recently, battered and fried. They are also incorporated into salads and stir frying. Lima bean pods are less palatable than snap bean pods. The seeds are eaten in the soft, immature form or cooked from their dry edible state. Almost all soft lima bean seeds are processed, mostly via freezing. Most dry beans are grown commercially in North Dakota, Michigan, Nebraska, Colorado, California, and Idaho.

Legumes, such as beans, contain phytohemagglutinin. Among the bean group, red and white kidney beans and scarlet runner beans contain high amounts of this toxin. Eating raw beans may cause nausea and vomiting. Boiling the beans for at least ten minutes reduces the toxin to safe levels. Beans should not be prepared in slow cookers, because their temperature is not high enough to reduce the toxin.

Table 16.2 Dry edible beans.

Common Name	Color	Other Descriptors	Flavor	Uses/Additional Information
Black (*Phaseolus vulgaris*)	black skin	dark cream to gray flesh; small, oval	mild, sweet, earthy; mushroom-like	also called turtle beans; Latin, Caribbean, Southwest American cooking; soups, stews, sauces; brownies and pizza
Blackeye (*Phaseolus vulgaris*)	white skin with small black eye	kidney shape; fine wrinkles; aromatic; creamy texture	earthy	African cooking; cooks rapidly; also called cowpea and black-eyed pea
Cranberry (*Phaseolus vulgaris*)	white skin with red markings	small, round; creamy texture	mild, somewhat nutty	red markings disappear with cooking; Mediterranean cooking; pizza

Garbanzo (*Phaseolus vulgaris*)	beige to pale yellow	creamy texture	nut-like	also called chickpea; Middle East and Indian cooking; salads, falafel
Great Northern (*Phaseolus vulgaris*)	white	medium size; flat; kidney shape; creamy texture	mild, delicate, somewhat nutty	adopts flavors of other foods; French cooking; Boston baked beans; pasta or pork
Kidney, Dark Red (*Phaseolus vulgaris*)	dark, reddish brown	kidney shape; soft texture	robust	adopts flavors of other foods; casseroles, salads, chili
Kidney, Light Red (*Phaseolus vulgaris*)	light red	large; kidney shape; soft texture	robust	adopts flavors of other foods; Caribbean, Portuguese, Spanish cooking; chili, salads, paired with rice
Lima, Baby (*Phaseolus lunatus*)	creamy white or green	flat	buttery	soups, stews, casseroles, plain
Lima, Large (*Phaseolus limensis*)	ivory	flat; creamy texture	buttery	American succotash; potato or rice substitute; soups, stews, casseroles, white bean chili
Navy (*Phaseolus vulgaris*)	white	small; oval; powdery texture	mild	adopts flavors of other foods; also called pea beans; pork and beans, baked beans, soups, stews, pureed, cookies
Nuna (*Phaseolus vulgaris*)	many colors, some mottled	thick-skinned;	peanut-like;	also called popping bean because they pop when cooked in oil and have consistency similar to popcorn; Peruvian and Ecuadorean foods
Pink (*Phaseolus vulgaris*)	pale pink	small; powdery texture	meaty	related to kidney beans; Western American cooking; chili
Pinto (*Phaseolus vulgaris*)	mottled beige and brown	medium; oval; powdery texture	earthy	related to kidney beans; turn brown when cooked; Mexican and Tex-Mex cooking; refried beans
Small Red (*Phaseolus vulgaris*)	dark red	smaller than red kidney;	robust	also called Mexican red beans; Creole cooking; soups, salads, chili

17

plants that add flavorful zing

plants that are used to flavor foods are called spices or herbs. Spice and herb use can be traced to ancient times, where many also served as important sources for medicinal treatments, dyes, and perfumes. In foods, spices and herbs enhance flavors present in the food or provide the main flavoring we sense through smell and taste. Although valued for their flavors and aromas, spices and herbs also have nutritional value in the human diet. The terms "herb" and "spice" are often used interchangeably but they are different.

Spices are derived from herbaceous and woody plants. All plant parts, except the leafy tissue, may be used as spices and include everything from roots to bark and flowers. Most plants used to produce spices grow in sub-tropical or tropical climates. The flavor and aroma of spices tend to be stronger in spices than in herbs, so smaller quantities of spices are usually used in cooking and making perfumes. Some herbaceous plants used for making spices may be easily grown in home gardens. Other plants that produce spices may require special conditions that would make them easier to grow in a greenhouse. Woody plants from which we obtain many spices would not tolerate the temperate climates affecting much of the United States.

Herbaceous plants generally have green stems and are not woody. "Herb" is a shortened form of "herbaceous" and, in this sense, refers to plants whose leafy tissues are used for culinary, medicinal, or perfumery purposes. Plants used as herbs most often thrive in temperate climates. Herbs are more flavorful when fresh, but many can be dried or frozen for later use. Most herbs are easy to raise either indoors in pots or outdoors in a garden or container. Sometimes, herbs are intermingled with other plants in a garden to ward off selected insect pests on the adjacent plants. So what is the main difference between the two? An herb usually comes from plant leaves and a spice comes from any other part of the plant.

Some perennial herbs may spread aggressively to the point they could be considered a weed. Before growing herbs, make sure you know their growth habit, so encroaching plants can be restricted to specific growing locations, like pots or outdoor plots surrounded by non-soil structures.

Occasionally, spices and herbs may be derived from the same plant. Seeds and leaves from the dill plant (*Anethum graveolens*) flavor pickles and prepared foods. The seeds are classified as a spice, while the leaves are herbs. *Coriandum sativum* is commonly referred to as coriander when its seeds are used as a spice and cilantro when its leaves are used as an herb. Some mustard species described below are harvested for the use of both leaves and seeds.

a. mustard and horseradish

Mustards (*Brassica* spp. and *Sinapis alba*) and horseradish (*Armoracia rusticana*) belong to the same family as the cole crops, the Brassicaceae. They are also native to Europe and western Asia. Like the cole crops, mustard and horseradish have strong, distinct, pungent flavors and aromas. Although most plants harvested for use as spices grow best in warm climates, mustard and horseradish are more suited for temperate climates like those throughout much of the United States.

Mustard

Several species of mustard exist, mostly in the *Brassica* genus, including field mustard (*B. rapa*), black mustard (*B. nigra*), brown and oriental mustard (*B. juncea*), yellow mustard (*B. hirta*), and white mustard (*Sinapis alba*). In Europe, yellow mustard is also called white mustard, leading to some confusion whether white mustard is the same plant as yellow mustard or a different genus and species.

Field mustard is the common name for the plant when used as an herb from its green leaves or as a spice from its seeds. Turnip is the common name when this plant is used for its root. Brown mustard leaves also are commonly used greens. Other mustard leaves are sometimes consumed as well.

Figure 17.1. Brassicaceous plants (plants in the mustard family) are very diverse but all of them produce small four-petaled flowers.

Canada and North Dakota account for most of the commercial mustard production for consumers in the United States. Most Canadian-produced mustard is exported to France, whose citizens consume the most mustard per person in the world. Yellow mustard is the primary species grown in the United States, followed distantly by brown and oriental mustard.

Mustards are herbaceous annuals that produce yellow, four-petaled flowers common among Brassicaceae plants. Flowering continues several days, with the greatest seed produced by flowers pollinated the earliest. Plant height varies from two and one half to nearly four feet. Their tap roots may extend five feet in suitable soils. Some mustard species spread so aggressively from the distribution of their seed that they are considered invasive or noxious and prohibited for growth in some states.

Mustards grow best in cool temperate climates. Due to diseases common to Brassicaceae species, fields should be avoided for planting other species in this family for three or four years. This process of crop rotation allows for diseases that impact these plants to diminish in the field over time. Loam soils with ample organic matter and pH 6.5 or greater produce the best crops. Mustards grown for their greens may be irrigated, although crops grown for their seed are usually not irrigated. Crops are established using transplants or direct seeding.

Mustard greens may be harvested by hand or machine. They should be marketed immediately, because they do not store well for long periods. Plants grown for their seed are harvested using combines when seed pods are dry, but before the pods split and seeds shatter.

Some condiments are spiced by mustard seeds. The seeds also flavor foods such as salads, soups, sauces, and pickles. White mustard seeds have the mildest flavor. Black mustard seeds produce the strongest flavor. Seeds are ground to make typical paste-like condiments and dry powder. An acid, such as vinegar, mixed with the ground mustard is the base component of mustard-containing condiments. Dry mustards season foods. Whole seeds are used when preparing pickles.

Yellow mustard seeds provide the flavor to the most popular condiment used on hot dogs in the United States. Turmeric (*Curcuma longa*) gives this mustard its characteristic bright yellow color. French-style, or brown, mustard condiments use a higher percentage of brown mustard seeds compared to yellow mustard seeds. Their flavor is more intense than yellow mustard condiments.

The inclusion of wine distinguishes Dijon mustard from yellow and brown mustard condiments. As a result, Dijon mustard has more calories than the other two mustards, albeit a much lower calorie count than ketchup condiment, which is sweetened with sugar.

Mustard seeds have high amounts of many vitamins and minerals, including vitamins A, C, K, and some B, and the minerals calcium, copper, iron, manganese, and selenium. Leaves contain similar vitamins and minerals found in seeds, plus the minerals phosphorus and potassium.

Horseradish

Horseradish is native to coastal regions of Europe. In the German language, this plant is known as "meerrettich," which translates to "sea radish" in English. The "horse" associated with the English name for this plant is speculated to have come from the "meer" in German sounding like "mare" in English. The strong flavor of horseradish root generally limits its use as a food to the spice category, rather than being served as a standalone vegetable.

Considered a minor crop, horseradish is commercially grown mainly in California, New Jersey, Virginia, Illinois, and Wisconsin. To ensure the best root development, soils should be slightly acidic, fertile, deep, well drained, and contain ample organic matter. Irrigation during dry periods improves crop yields.

Horseradish is usually established from root cuttings planted by machine or hand. Modified potato harvesting equipment may be used to harvest horseradish roots. If not processed immediately, roots should be stored in cool, dark environments. Horseradish roots harvested latest in the fall store better than roots harvested before cool temperatures have arrived.

Horseradish roots are grated for use as a spice because of their biting flavor. The grated material loses its strength fairly soon after exposure to air or heat. Unless used immediately, this spice is mixed with vinegar to preserve as much of its flavor as possible. Meats are most commonly flavored with horseradish. Soups, sauces, salads, and sandwich spreads also contain this pungent spice. Nutrient value is lower than many other spices and herbs, although horseradish has some vitamin C and moderate amounts of minerals.

b. mints—more than just flavors of gum

The mint family (Lamiaceae) includes hundreds of mint varieties, although most are not sold commercially. Plants in this family are known for their square stems, making them easily identifiable. Mints add invigorating flavors to many foods. Each variety and species has their own subtle flavor and scent distinguishing them from other mints. Peppermint (*Mentha x piperita*) and spearmint (*M. spicata*) are among the most popular cultivated mints.

Figure 17.2. Not only are the flowers of plants in the mint family unique, but every mint has a square stem.

Many mints grow so well without human intervention that they may become overpowering in the landscape. Moist soils encourage the most vigorous growth. Mint plants spread by horizontal stems, which can be used to propagate new plants. Stem cuttings are another method of vegetative propagation.

Peppermint is a hybrid formed from a cross between water mint (*M. aquatica*) and spearmint. Although most mints have green leaves, chocolate mint, a variant of peppermint, has dark reddish-brown leaves. The plants have a slight chocolate aroma and the leaves have a mild, minty chocolate flavor. Chocolate mint shoots were observed more than 20 feet from an original planting within a few years. The new plants even survived unmanaged areas in rocky ditches with poor soil.

Both peppermint and spearmint have non-food and food-enhancing uses. These herbs are incorporated into recipes and also serve as garnishes for foods and drinks. Peppermint and spearmint contain protein, vitamins A, C, and B-complex, and several minerals, including calcium, copper, iron, magnesium, manganese, phosphorus, potassium, and zinc.

The breath-freshening effect of peppermint and spearmint is incorporated into chewing gum, toothpaste, mouthwash, breath sprays, and lozenges. Peppermint may relieve digestive upset, although anyone with acid reflux should avoid this herb because it may exacerbate that condition.

The feline species of animals has its preferred flavor of mint, appropriately named catnip mint (*Nepta cataria*). The strong odor of this mint has been observed to exhilarate some cats and make other cats rather mellow. Humans use the leaves in salads and for flavoring some meats.

c. parsley, sage, rosemary, and thyme

A refrain from the 1960's song "Scarborough Fair," adapted from a centuries-old English ballad, highlights the interesting combination of herbs in this section. Each herb is popular with cooks and infuses its own flavor to foods.

Figure 17.3. Parsley is typically used as a garnish, but can also be eaten.

The carrot family (Apiaceae) includes several plants utilized in food preparation. In addition to parsley (*Petroselinum crispum*), the herbs cilantro, dill, and fennel (*Foeniculum vulgare*) are classified in this family.

Not all plants in the Lamiaceae (mint) family smell or taste "minty" like peppermint or spearmint. Examples include the herbs sage (*Salvia officinalis*), rosemary (*Rosemaryinus officinalis*), and common thyme (*Thymus vulgaris*).

Parsley

Parsley is most recognizable by it divided leaves. Leaf blades on Italian types of parsley are flat. Other cultivars, usually called common parsley, have slightly curly to strongly curly, bright green blades. This biennial species is grown as an annual in most gardens.

Seeds are easy to grow, although slow to germinate. Starting plants indoors while it is too cool to plant them outside helps the parsley reach harvestable size earlier in the season than when they are sown outdoors from seeds. This herb may also be grown completely indoors in pots.

Optimal growing conditions include full sun and soil rich in organic matter that also drains well and has a pH slightly acidic to neutral. Parsley grows best with moderate fertility and sufficient moisture to prevent wilting. When grown indoors, locate the plants in the window with the most sunshine and make sure the pots have openings in their bottom to allow adequate drainage.

Parsley does not have as distinct a flavor as many other herbs. It is commonly used fresh to adorn restaurant plates. Although many people do not consume this herb when presented as a garnish, parsley contains vitamins A and C, iron, and sulfur. Eating parsley may stimulate appetites before a meal. After meals, consuming the parsley garnish cleanses one's breath.

Sage

Several varieties of sage are available. Leaves of this evergreen, shrub-like perennial species are a distinct pale grayish-green color. Varieties with variegated purple and yellow leaves do not have as strong a flavor as the non-variegated varieties. They are more commonly used for their ornamental value. Sage has distinctive purple flower spikes. Plants may reach two feet in height.

Some sage varieties must be propagated by cuttings, while seeds or cuttings are suitable for other varieties. Sage may be grown outdoors in gardens or containers.

This herb is easier to grow in cooler regions of the temperate zone. Sage tolerates cold, being hardy enough to survive northern winters with protection from mulches. To keep the most tender, edible leaves available, some experts recommend replacing the plants every few years.

Soils should be well-drained without excess clay. Sage plants do best at a slightly acidic pH. Once established, plants are drought tolerant. Sages grow best in full sun exposure.

In addition to leaves, sage blooms also are edible. This herb is often associated with poultry recipes. It is a favorite herb in Thanksgiving stuffing. Breads, cheeses, and other meats include sage in their recipes. Cooked vegetables may be enhanced by the addition of sage. Leaves can be used to make tea. The fruity flavor of pineapple sage (*S. rutilans*) complements desserts and drinks. Nutritionally, this species is high in many B vitamins and a good source of vitamin A and beta carotene.

Rosemary

Rosemary includes a diversity of varieties. The evergreen shrubs range from prostrate plants less than one foot tall to some that reach six feet in height. Plants may survive several years, making them desirable for ornamental uses including hedges. When harvested for its leaves, replacements are recommended every six years or less.

Propagation from cuttings or layering is preferred over starting plants from seed. Softwood and hardwood cuttings can be utilized.

Fertile, well-drained soils provide the best growing media. Rosemary is native to the Mediterranean region and has some frost hardiness. Plants may survive slightly colder climates if protected during winters; however, in most temperate regions a cold winter will kill them. Rosemary is most vigorous when grown in full sun.

This herb is used in the preparation of many meat cuts. Rosemary also complements other herbs, like sage, to flavor prepared foods. Whole cuttings used during barbecuing and in wood-burning stoves add a pleasant aroma to the atmosphere. Nutritionally, rosemary contains good amounts of vitamins A, C, and some B vitamins. It is also high in calcium, iron, and manganese.

Thyme

In addition to using its leaves in food preparation, thyme is valued as a landscape ornamental and for the production of bonsai creations. The evergreen plant species of the *Thymus* genus thrive in a range of temperate climates around the globe. Plant selection based on cold tolerance greatly increases successful establishment of thyme species. Where the climate is too cold for year-round growth, thyme may be grown outdoors as an annual or indoors in pots.

The tallest thyme species do not usually extend vegetation more than one foot above the soil. Some species are approximately one inch tall, making them suitable for use as groundcovers. When stepped on, leaves of thyme used as a groundcover release a pleasant scent. Between stones on a pathway, the low-growing varieties of this species are protected from extensive damage, while tolerating some leaf-crushing that allows aromatic oils into the atmosphere.

Vegetative propagation by softwood cuttings is most common. Some plants also are started by layering. Seed is seldom used, since there may be excessive variation among the seedlings.

A somewhat stressed environment produces the most flavorful thyme leaves. Plants grow readily in many soil conditions having minimal fertility and good drainage. Thyme has good drought tolerance and thrives in full sun.

The evergreen nature of thyme provides leaves that may be harvested throughout the year. When plants become woody, their leaves may not be as flavorful. Replacement every few years is recommended for plants used as leaf sources for flavoring foods.

Thyme is popular in flavoring many foods associated with Mediterranean cuisine. This herb contains several vitamins (A, C, E, and some B vitamins) and minerals (calcium, copper, iron, magnesium, manganese, and zinc). Besides its food value, thyme has antiseptic value, making it useful in toothpastes and mouthwashes.

d. basil—it's the pesto

Perhaps best known as the herb giving pesto its distinctive flavor, sweet basil (*Ocimum basilicum*) is an annual herb that can reach three feet in height. Cooks grow this most popular fresh-cut herb throughout the year indoors in pots, as well as outdoors in containers and gardens during summers.

Basil can be established from seed, whether grown indoors or outdoors. Successive plantings provide a regular supply of tender leaves. To prevent leaves from becoming bitter, flower buds should be removed as soon as they appear. Nutrient-rich loam soils favor basil development. Cool soil and air temperatures delay growth.

Sweet basil, a favorite seasoning in Italian cooking, enhances sauces, salads, and pasta. This herb may be added fresh, dried, or combined with pine nuts and olive oil to make pesto. While best known for its use on pasta, basil also has been added to many foods, including pizzas and sandwiches.

Other basil species, used in lesser quantities, add unique aromas in herb gardens, have non-food uses, and/or add flavor to prepared foods. For a strong lemon scent in herb gardens, grow lemon basil (*O. americanum*).

Annual shrubs in the *Ocimum* genus include camphor basil (*O. kilmandscharicum*) and holy basil (*O. tenuiflorum; O. sanctum*). Camphor basil finds its way into closets to help prevent moth damage to woolen materials and is used in making teas for soothing stomachaches. Holy basil is considered sacred in the Hindu religion. This species of basil has both medicinal and food-preparation value.

Like many other herbs, basil has good nutritional components. This herb contains protein and vitamins A, C, E, K, and B-complexes. The essential minerals present include calcium, copper, iron, magnesium, manganese, phosphorus, potassium, and zinc.

e. bulbs we eat—onions, shallots, garlic, and leeks

Several underground storage organs of related plants in the *Allium* genus of the monocotyledonous Liliaceae family may be used alone or as ingredients in several food preparations. Plants in this genus are usually classified as bulbs, because the underground portions of the plants have several leaves surrounding a compressed stem. The outer leaves, or skins, are papery. The skins protect the internal succulent leaves, which provide the food for the following season's plant. Adventitious roots grow from the stem's base.

Among the most popular bulbs humans consume are onions (*Allium cepa*) and garlic (*A. sativum*). Shallots (*A. cepa*, Aggregatum group) and leeks (*A. porrum*) do not form distinct bulbs, but we eat their leaf tissues, which are similar to onions. Onions and leeks are biennials grown as annuals. Garlic and shallots are perennials, but are usually harvested as annuals. All these species are believed to originate from areas in Asia.

Onions

Onion skins may be yellow, brown, purple, or red. Most internal bulb leaves are white to greenish, with some having purple hues on the outer sides of each leaf. Onion cultivars vary from mild to strong flavors. Non-American onion types are milder than American types. The two major non-American types are Sweet Spanish and Bermuda onions.

When onions are cooked until brown, or caramelized, most have a sweet component. Some cultivars grown in specific locations taste somewhat sweet in their raw form. These include Vidalia onions from Georgia and Walla Walla onions from Washington State. Due to their tender leaves, these onions are consumed fresh. Chefs frequently use them for making onion rings.

Most commercially grown onions in the United States are planted from seed. Less frequently, transplants

Figure 17.4. Onions have been domesticated for thousands of years having been prominent in Egyptian, Greek and Roman civilizations.

or sets are utilized. Transplants are established from seed earlier than field-planted seeds. This allows the transplanted onions to mature earlier than seeded onions. A set is a group of small plants that look like slightly dried bunching onions. The plants are produced during the prior growing season. Sets are easier to plant and also mature earlier than seeds. Home gardeners mostly use onion sets or transplants.

The best crop yields occur on muck soils. Muck soils are composed of heavily decomposed organic material and are strikingly black, derived from drained swamplands. Although not extremely common, these soils do occur in pockets worldwide and the largest contiguous acreage is thought to be in New York State. Onions require ample fertilization and irrigation to develop fully.

American and Sweet Spanish onions are planted in the spring in more northern sections of the United States and harvested in the fall as full-size bulbs. Bermuda onions are usually grown in southern states, where they are planted in the fall and harvested the following spring. California produces the most onions in the United States.

Spacing depends on mature bulb size and shape. Shapes include globe types and variations of globe types, as well as spindle, flat, flat-top, and top types. Planting depth influences bulb size. Shallow bulbs are usually somewhat flattened. Bulbs planted more deeply attain their characteristic shape, depending on cultivar. Numerous plant pests make onions an expensive crop to produce. Chemical controls are often used extensively.

Some onions, known as bunching onions for the way they are packaged for sale, are harvested early in the growing season. Green bunching onions are slender, with only slight bulb formation. Bunching onions should be consumed fresh.

Aboveground leaves should be withered and bent over before harvesting larger, mature bulbs. Plants may be pulled by hand or undercut by machines and lifted. Tops are cut back to roughly two inches above the bulb and roots are trimmed. Before storing, onion bulbs must be cured and dried in the field or using forced air drying. Curing seals off the cut tops and drying helps prevent disease formation.

Cultivars vary in storability. American onion types are the predominant type produced in America because they store better than non-American types. Onions destined for storage may be treated with a chemical to prevent sprouting.

Early uses of onion included aiding the mummification process, curing baldness, and as an antiseptic. Today, onions are primarily consumed alone or in combination with numerous foods. Onions are eaten cooked or raw. Chopped, sliced, and diced onions add flavor to many recipes, including salads and casseroles. They are popular on cooked meats like hamburgers and hot dogs. Onions may also be pickled alone or with other vegetables like cucumbers. In the human diet, onions provide fiber, vitamins B6 and C, folate, potassium, and manganese.

A popular vegetable combination is pearl onions and peas. Pearl onions (*A. ampeloprasum* var. *sectivum*) are roughly an inch or less in diameter. Despite their name, pearl onions are more closely related to leeks than the common onion.

Shallots

Shallots are small onions that do not form the pronounced bulb shape like other onions. Significantly less acreage is planted in shallots compared to the other crops in this section. This perennial species matures early, producing several new propagules each year. The clustered shallot bulblets are separated for planting the next season's crop.

Shallot production is similar to green bunching onion production. They are harvested while the foliage is green and eaten immediately, with none put into storage. Both green leaves and bulbs are consumed, usually as a flavoring component of prepared foods. Nutritional value of shallots includes vitamins A, B6, and C, folate, manganese, and potassium.

Garlic

Garlic bulbs grow differently than onion bulbs. Garlic bulbs are segmented. Each segment is called a clove. Each clove contains a bud capable of producing a plant. The strongest-flavored of the species described in this section, garlic was once believed to be able to ward off witches and vampires. Today, garlic is used most often to enhance flavors of prepared foods. Fresh garlic is usually incorporated using whole or chopped cloves. Dehydrated garlic is processed into powder or flakes.

Although California is the most prolific producer of garlic, its acreage is minor compared to onion production in that state. Climates that favor onions also favor garlic. The separated cloves are utilized for planting. While onions grow best in muck soils, mineral soils with more sand are preferred for garlic production. Bulb

Figure 17.5. While onions and garlic appear similar in the field, onions have a single large bulb while garlic is composed of numerous smaller bulbs or cloves, attached to each other.

formation may be adversely affected when there is excess or insufficient moisture for growth or soils with high clay content are used. Other production practices for garlic are similar to onion procedures.

This bulb has antibacterial, antifungal, and antiviral abilities. Research has suggested garlic may relieve digestive disorders and reduce cholesterol, heart attack risk, and incidence of cancer. Nutritionally, garlic is a good source of vitamins C and B6 and the minerals calcium, manganese, phosphorus, and selenium.

Leeks

We eat the blanched leaf bases of leeks and while they are included in this section of bulb-producing plants, leeks do not actually form a true bulb. Planting, harvesting, and post-harvest handling of leeks is similar to onion production. The depth of planting determines the amount of the desired white tissue. Some growers also mound additional soil around the growing plants to increase the length of blanched material. Plants are harvested when diameters of blanched leaves are between one half and two inches.

Leeks have a similar, but milder, flavor compared to onions. They are prepared raw or cooked. Leeks are more nutritious than onions, containing high levels of potassium and iron. Leeks also have vitamin C and some beta carotene, the precursor of vitamin A.

a. corn

corn (*Zea mays*) is one of several annual plants whose seeds are commonly referred to as grains. Grains can then be divided into large and small grains. Corn is a large grain, while wheat, oat, barley, and rice are small grains. Corn grains are also referred to as kernels. The grain crops are all monocots in the Poaceae family, which is sometimes referred to as the grass family. Corn is a grass, just like rice, wheat, oats, and barley!

The middle section of the United States, where corn is grown on the same land each year, is referred to as the Corn Belt. One farm may grow thousands of acres of corn each year. On smaller farms, corn is more likely to be part of a rotation of crops that also include small grains and legumes. Where corn is continually grown, significant nutrients are removed from the soil every year. This necessitates large fertilizer inputs each year to produce optimal yields. While the soil is bare between crops, soil erosion may be significant.

Corn is grown so widely throughout the United States because it is an extremely useful plant. It can be processed many ways and plays a role in the production of many different commodities. High fructose corn syrup is an ingredient in many foods and comes from corn. Corn oil, pressed from the seed embryos, is also common in things like margarine and a myriad of products, from soap to ink and pesticides. Corn is the primary constituent of bourbon, and cornstarch and corn flower are used in many foods. Americans consume more corn than any other product.

The ancestor of corn may be a smaller plant known as teosinte, although other ancestors may also exist. Teosinte produces many small, slender ears, some less than one inch in length, whose seeds shatter upon maturity, allowing them to be distributed wide and far. Teosinte is native from Mexico to South America.

There are many types of corn, although they may all look similar. Field corn (*Zea mays* var. *indentata*) is also called dent corn. As the ears mature, the hard starch to the outside of the kernel remains firm, while the soft starch to the inside is not as firm and shrinks, resulting in the center of the kernel becoming sunken, or dented. Most of this corn is used for animal feed, although a significant portion of it is now used to produce ethanol for fuel in cars and other vehicles through fermentation. Most gasoline has at least 10 percent ethanol mixed into it, which is produced from corn.

Sweet corn (*Zea mays* var. *saccharata*) is different from other corn because it is consumed while still immature. Most varieties have been bred to have a very high sugar content; the ears contain some white and some yellow kernels. Popcorn (*Zea mays* var. *praecox*) was used by Native Americans prior to the arrival of European settlers. The hard seeds appear more teardrop-shaped than the conventional kernel of other corn varieties. Moisture in the middle of the kernel heats up as the popcorn is cooking and eventually gets hot enough to make the kernel burst and literally turn inside out. Popcorn is very high in fiber.

An alternate use of field corn is the production of corn silage for livestock feed. Corn is harvested near the end of the growing season before the ears are fully mature and the stalks still have mostly green color. The entire part of the plant above the soil is harvested and chopped into coarse pieces prior to placement in an enclosure.

Silage is produced when the plant material ferments, similar to the process of making sauerkraut. As the weight of the corn presses down on lower material, liquid is expelled from the vegetation and the green color changes mostly to a greenish yellow color, with the silage producing a strong odor as a result of the fermentation process. In addition to conventional silos, silage may also be produced in horizontal bunkers using plastic covers.

All varieties of corn are monoecious (they have separate male and female flowers). The female flower eventually becomes an ear of corn, while the male flower is the tassel at the top of the plant. Pollen from the tassel moves by wind until it comes in contact with the stigmas of female flowers on the same or different plants. The part of the corn ear usually referred to as the silk is actually the stigma and the extremely long style of the female flower. The ovary attached to each silk produces one kernel of corn when fertilized.

Some corn plants may be ten or more feet tall. In order to help support the stems, special roots grow adventitiously from the stem to the soil. These roots, known as prop roots, are much thicker than the roots in the soil.

Figure 18.1. Corn is one of the most important field crops in the United States, grown nationwide and used to produce thousands of different products from food to fuel.

b. soybeans

Unlike corn and wheat, soybean is a legume (they belong to the Fabaceae family, formerly known as the Leguminosae family). Common legumes include peas, clover, peanuts, locust trees, and even wisteria. Legumes form an association with a bacterium (*Rhizobia* spp.) that converts nitrogen in soil air pockets into a form that plants can utilize. In return, the plant provides carbohydrates for the bacteria. Legumes can be identified by

Figure 18.2. Soybeans belong to the same family as peas, the Fabaceae. Pea flowers are easily identified by their shape, typically described as similar to a boat's keel.

examining their flowers (all look very similar to pea flowers) or by the root system, where small bumps or nodules can be found. These nodules contain the *Rhizobia* bacteria. Legumes are also unique because they contain very high levels of protein, which is the primary reason why soybeans are so useful.

Legumes have a tremendous advantage over most other plants. Because they have nitrogen-fixing bacterial nodules, they are much less dependent upon nitrogen fertilizer than most other plant species. As a result, when we grow legumes agriculturally, we don't need to apply as much fertilizer. When these plants grow in nature, they are often very successful in nutrient-poor soils because the bacteria provide them with a built-in fertilization system.

Legumes can even provide nitrogen for the plants growing next to them. The next time you see a lawn full of clover patches, take a careful look at the color of the patches. If the lawn is poorly maintained, those patches will be bright green because of the nitrogen from the bacteria. The other areas of the lawn will be weak or off-color.

Soybeans (*Glycine max*) originated in East Asia, but are grown widely through the northern regions of the Midwestern United States, as far north as Minnesota. More soybeans are grown in the United States than any other country. Soybeans have many uses, including vegetable oil, soy milk, tofu, and a number of fermented products like soy sauce. In addition, because soybeans are so high in protein, they are often used in animal feed. Much like corn, they can be commercially processed and are incorporated into many different foods, medicines, cosmetics, and industrial products.

One of the most significant expenses in growing any crop on a large industrial scale is the cost of weed control. In many cases, weeds need to be managed through the use of mechanical equipment that physically disturbs the weeds growing within the actual crop, or the weeds need to be picked manually. This is all very time-consuming and expensive. If weeds are not controlled, the weeds steal water and fertilizer from the crop and crop yield suffers. In the past 20 years, many different crops have been genetically engineered to be resistant to the nonselective herbicide glyphosate (known as Round-Up), including corn, soybeans, and cotton. When a Round-Up–ready crop is grown, farmers can apply glyphosate to the entire field a few times a year, completely controlling the weeds while the crop remains resistant to the chemical. When this practice is undertaken on soybeans, the total number of pesticides used on the crop is reduced.

There are a few drawbacks to Round-Up–ready crops. Firstly, many countries have historically prohibited the import of genetically modified crops. Genetically-modified crops are those that have had genes inserted into them from related or unrelated species. In addition to this issue, the use of so much glyphosate is not likely to be a permanent solution—eventually, the weed populations will become resistant to the chemical also.

Finally, seed production practices are affected by Round-Up–ready crops. Until recent years, farmers would save some of the soybeans they grew each year so they could reseed their fields the next year. The introduction of genetically-modified soybeans resistant to the nonselective herbicide glyphosate (Round-Up) requires new seed to be purchased each year, which is an expense growers did not have in the past.

Figure 18.3. Soybeans come in many varieties and are a major crop in the United States, using in many different food products and processed foods.

c. wheat

Wheat has often been referred to as the staff of life because it is a staple (or primary food) in the diets of so many people worldwide. Evidence of the use of wheat by humans dates back at least 12,000 years. In fact, wheat was such an important crop for early civilizations that the failure of repeated wheat crops could lead to the collapse of cities and entire nations. Wheat is currently grown on more land throughout the planet than any other crop, but it originated in the area between Northern Africa and Turkey. The most common use for wheat is flour that can be made into bread and related products. It can be made into pasta, crackers, cookies, muffins, doughnuts, breakfast cereals, and beers. It can also be fed to animals.

Wheat was one of the first agricultural crops to be domesticated and cultivated. Early civilizations began growing wild types of wheat, and over time, they collected seeds from the plants that produced the largest and most seeds. Over hundreds of thousands of years, wheat plants were improved through this process. These plants were highly adapted to the environments they were used in. Such plants are called landraces (locally produced strains initially used when crops were domesticated).

Figure 18.4. From the Middle East to the Midwest, wheat is one of the most widely grown crops in the world.

Modern-day breeding often uses some of the same techniques, in addition to many others, to produce the highest-yielding wheat varieties possible. The genus for wheat is *Triticum. Triticum aestivum* is common wheat, but many species exist. Although most wheat is planted in the spring, winter wheat is planted in the fall, survives the winter, and is harvested the following year in early summer.

A worldwide breeding effort to produce cultivars suitable for use in third world countries was later termed the Green Revolution. One of the key breeders in this effort, Norman Borlaug, was awarded the Nobel Peace Prize for working to improve the life of disadvantaged people throughout the world. He contributed a semidwarf wheat at the International Maize and Wheat Improvement Center in Mexico City, Mexico, CIMMYT (Centro Internacional de Mejoramiento de Maíz y Trigo). Borlaug also worked on breeding wheat and rye to develop a hybrid. The product of this effort was known as triticale, a combination of part of each of the plants' genus names (*Triticum*) for wheat and (*Secale*) for rye.

The cultivars developed as part of the Green Revolution had some difficulties. The best yields required adequate available water, fertilization, and pest control. These inputs were often too expensive for subsistence farmers to afford (subsistence farmers live off the food they grow, but do not generally sell much of it). The result was that the most frequent users were operations with sufficient monetary resources to afford the inputs. This dilemma continues today, particularly for growers who attempt to manage crops sustainably. Sustainable management requires growers to use limited numbers of inputs. Unfortunately, many of the commercially available varieties were developed for high-input management, regardless of whether the crop in question is wheat, corn, or tomatoes.

d. rice

This small grain is a key food source for many groups of people across the globe, with almost as much total production as corn. In Asia, where rice is believed to have originated, rice (*Oryza sativa*) is the primary grain in people's diets. Rice is high in starch and more nutritious when not processed to remove the outer layers. Intact rice is light brown, while processed rice, which most Americans are familiar with, is white.

The ancestor of today's rice (*Oryza rufipogon*) still exists in Asia and is called wild rice, not to be confused with the wild rice (*Zizania spp.*) native to the United States. The wild rice from Asia will cross (or mate) with the *O. sativa*. Both grow in the same location, have similar genetic material, and similar growth habits that bolster the contention that *O. rufipogon* is an ancestor of *O. sativa*. One significant

Figure 18.5. While some rice is grown in the Unites States, the vast majority of the rice crop comes from China and Southeast Asia.

difference between the two species is that the ancestor is perennial, while the modern species lives as an annual.

Rice native to Asia has two subspecies used by most humans. The Japonica subspecies grows best in a temperate climate. Optimal yields occur when Japonica is managed for exact amounts of water, fertilization, and timely insect control. It does not thrive in humid climates. Contrasted to the Indica subspecies, Japonica has shorter stalks, matures earlier, and yields well when grown under specified conditions. Its seeds are shorter than Indica subspecies, so they are referred to as short-grain rice. Upon cooking, Japonica becomes stickier than Indica, whose grains remain separated.

The Indica subspecies grows in humid areas, including those where monsoons regularly occur. Indica plants are taller and leafier than Japonica plants. Fertilization does not improve grain yields as it does with Japonica. Indica's taller stalks make it more susceptible to lodging, a condition where stems fall over and make harvesting more difficult. The longer seeds of Indica are called long-grain rice.

Much of the rice produced today in Asia is still planted and harvested by hand. For paddy rice, seedlings are placed into the water-covered soil. The water aids in weed control. Some rice is planted away from continually water-covered land. This rice is known as upland rice, while rice planted underwater is named lowland rice. Upland rice is irrigated as needed, but the plants are not grown in paddies and it is very drought tolerant.

In this country, rice production occurs in southern regions, with Texas being the leading producer, but it is also grown in Arkansas, Louisiana, and California. Most rice produced in this country uses the upland method.

As part of the Green Revolution, rice improvement was sought. The International Rice Research Institute developed the cultivar IR72. The breeding of this cultivar included many crosses involving 22 landraces . IR72 is a semidwarf variety to reduce lodging, producing higher yield than previous cultivars, with increased grain yield when fertilized with nitrogen.

An unrelated genus includes the grain we commonly call wild rice (*Zizania palustris*). This is not the wild rice known in Asia. Native Americans have harvested this rice for hundreds of years. Wild rice in the United States grows naturally in waterways in the northern states, including Minnesota. During fall, after the wild rice has matured, the rice is harvested in much the same way it has always been done. People in boats harvest stems laden with ripe grains by bending the stems into the boat and beating the grains into the boat. Due to shattering of wild rice seed, some seed falls back into the water, necessary for producing the following year's crop. The limited availability of wild race causes this grain to be more expensive than white or brown rice. Six dollars per pound is a common price for wild rice in retail grocery stores.

Cultivated wild rice cultivars are being grown in both Minnesota and California, but it is questionable whether we should still call any of these cultivars "wild rice" if they are produced as a managed crop.

The dark brown grain of wild rice splits open during cooking. Cooking time varies, depending on how crunchy or soft people prefer this grain. It has a distinct nutty flavor, making it desirable as an ingredient in numerous recipes. The high cost of wild rice results in its use as a flavoring ingredient, rather than the main ingredient. Small amounts of wild rice added to white or brown rice are sold for use as a side dish in meals.

e. other small grains (oat, barley, rye) and alfalfa

On a much smaller scale—but still important—are several small grains, including oats and rye. The legume alfalfa is not a grain, but it is an important crop used as animal forage and is often grown similarly to the small grains. Oats (*Avena sativa*) are often consumed by people as oatmeal or rolled oats, but are increasingly being used

Figure 18.6. Alfalfa is grown for many purposes including animal forage, cover crops and edible alfalfa sprouts.

in whole-grain breads and other products. The oat has gained attention in recent years as an important health food, but it is primarily consumed by animals, particularly horses and chickens. Horses are typically fed whole grains, while cows consume it in a ground feed that includes corn, oat, and soybean meal. Oats are occasionally also used in brewing beer.

Barley (*Hordeum vulgare*) is also a part of many human diets. Almost as ancient as wheat in the human diet, barley can be processed into bread flour and be used to ferment beer (just like wheat). In addition to being the most common source of malt for beer production, barley is a very common grain used in the production/fermentation of whiskey. Barley can also be incorporated into foods as a whole grain. Barley is well-suited to temperate climates and grows well in cool conditions, allowing its production in many different areas of the world where many other crops may not be as adapted.

Rye (*Secale cereale*) is closely related to both barley and wheat (but not ryegrass, *Lolium perenne*, which is used for turf) and can serve many of the same purposes. Rye is not as commonly planted as those two crops; however, it does have a better ability to thrive in poor soils. Rye is commonly used in many locations as a cover crop—that is, it is planted in the fall to keep soil erosion to a minimum and can improve the quality of the soil during the fall and winter when other crops would go dormant or die.

When mature, all the small grains are various shades of tan to golden brown. Combines sever stems close to the ground. The stems then move up a short elevator to a device that separates the stems from the grains. The grains are collected in a bin, while the stems are returned to the field. When the combine's bin is full, the grain is transferred to wagons or trucks that carry the grain to storage or processing facilities. The stems are often raked into rows and baled together for use as animal bedding. At this point, the stems are now called straw.

Unlike the small-grain grasses, alfalfa *(Medicago sativa)* is a legume. Like soybeans, it utilizes nitrogen-fixing bacteria in its root system to increase the level of nitrogen it has access to, which allows it to grow well in poor soils. Alfalfa is primarily used as forage—that is, as food for grazing animals. Alfalfa also differs from small grains because it is a perennial plant and can be maintained from year to year in the same field. Alfalfa fields are harvested multiple times each year, sometimes fed fresh to horses and other animals, but usually dried and baled to be fed as hay or fermented in silos or bunkers to be fed as haylage. Eating too much fresh alfalfa can cause bloating in ruminant animals like cows, so they eat alfalfa as hay or haylage. Although most alfalfa is grown in the Northern United States, it is also found in large amounts in Canada, Europe, and the Middle East. Unlike the other small grains, it is not directly consumed by people (unless in the form of alfalfa sprouts) but is frequently planted as a long- or short-term cover crop and a soil-stabilization crop.

19

a. cotton

cotton is one of the most important nonfood crops grown in the world today. If you look at the tags on most of the clothes you are wearing, you will likely find a lot of it contains some percentage of cotton. Because cotton can be processed into a textile (a cloth), it can be used for making clothes, towels, sheets, curtains, upholstery, and a myriad of other products. Cotton has many advantages as a textile material: It is comfortable to wear, it is rugged (denim jeans are all cotton), it is relatively easy to grow, it is easy to process industrially, and it is cheap, compared to other textile materials. Cotton requires a long, warm growing season, so production occurs in the Southern portion of the United States. Most of the cotton grown in this country comes from Texas, but most of the Southern states produce some level of cotton, including California. The majority of the cotton in the world is produced in China, with India close behind.

The scientific name for the cotton plant is *Gossypium hirsutum*. It is an annual plant belonging to the mallow family (the Malvaceae). The part we use commercially, the "cotton," comes from flowers. After cotton plants flower, seeds are produced, and the seeds are covered in hairs that grow from epidermal cells on the seed coat. These form a white ball (called a boll); each boll will contain many thousands of hairs or fibers. The bolls are processed to remove the seeds and other contaminants and then spun into lengths of cotton thread. The seeds can be saved and pressed for vegetable oil.

The production of cotton has greatly influenced the course of history, particularly in the United States. Initially, cotton was harvested by hand, often using slaves forced into captivity and taken from Africa. The process of hand harvesting was extremely slow and tedious, so slaves were forced to do it. As cotton production increased, more labor was required, and plantation owners needed more slaves to cheaply produce cotton. Cotton was one of the driving forces behind the growth of slavery in the Southern states. Cotton was also a useful crop in the South because it is nonperishable. As long as it was dry, it could be stored for months or years before being sold.

Figure 19.1. Cotton is produced from the seeds of the cotton plant and harvested as bolls for processing.

At one time, half of all chemicals sprayed on agricultural crops were sprayed on cotton. In addition to pest and disease control, the use of chemicals—called defoliants—make the cotton plants drop their leaves at harvest time. Removing leaves makes harvesting cotton bolls significantly easier. Some pesticide use on cotton has been reduced with the introduction of genetically modified cotton, particularly varieties that are resistant to attack by the cotton bollworm.

Another species of cotton, *Gossypium barbadense*, is perennial and can be grown a number of years in the same field (it's really a small tree or bush). This cotton is called "pima" or "Egyptian" cotton and produces fibers that are extra-long, very strong, and very soft. Most of the cotton grown in Arizona is of this type.

b. tobacco

Tobacco plants are typically large, leafy annuals that belong to the genus *Nicotiana*. There are many different species and varieties of tobacco grown through the world, with *Nicotiana tabacum* being the primary species. Although tobacco can be used medicinally and even as a pesticide, the primary use for tobacco is consumption in the form of cigarettes, cigars, pipes, and chewing tobacco. The tobacco plant is a member of the *Solanaceae* family, just like tomatoes and potatoes. As a result, it produces alkaloids. Unlike many Solanaceous plants, it only produces nicotine and no other toxins (many alkaloids are highly toxic). Nicotine is primarily addictive in humans and not highly toxic, but it can be poisonous to insects, hence its use as a pesticide.

Tobacco production has supported the economy of many Southern states and is currently widespread throughout North Carolina, Virginia, Tennessee, and Kentucky. Despite the causal link between tobacco products and cancer, tobacco continues to be grown on much of the same land that has been in cultivation for tobacco farming for decades. Tobacco generates significant revenue for growers in a relatively stable national market (tobacco consumption is actually growing internationally). Unfortunately, successful tobacco production requires large amounts of pesticides, which—even at legally acceptable levels—are directly consumed by smokers.

Although usually thought of as a Southern crop, tobacco has been grown in New England since about 1643, particularly in the Connecticut River Valley, where high-end cigar wrapper tobacco is still grown at a low level today. Unfortunately, the disease known as blue mold is common in this region (caused by *Peronospora tabacina*). Because cigar wrapper tobacco is expected to be aesthetically perfect, even a small amount of this pathogen and others can cause a crop failure. While the pathogen is still a problem in Southern states, slightly damaged tobacco (up to 25 percent leaf damage) from the South can still be used in cigarettes and

Figure 19.2. Tobacco farming is common in the Southeast United States but suburban encroachment continues to reduce the availability of land for farming.

Figure 19.3. Tobacco flowers typically range in color from light pink to purple and produce many extremely small seeds.

pipe tobacco. One of the other complications of growing tobacco is that it needs to be "cured," or dried, before further use. This process can be time-consuming and require large barns. Most other consumer crops do not require this type of practice.

In greenhouse production operations, the use of tobacco is strictly forbidden, due to the possibility of the introduction of tobacco mosaic virus (TMV) to other plants. TMV is difficult to control and can spread to many different plants. People who smoke must take care to thoroughly wash their hands before working with any greenhouse vegetation because the virus can actually be transmitted from a cigarette to the smoker's hands and then to the plants in the greenhouse.

c. flowers

All flowering plants are classified as angiosperms, including plants we do not usually categorize as flowers. These include many deciduous trees, shrubs, and vines. Plants grown primarily for the appearance and/or fragrance of their floral structures are the types most people would place in the category of "flowers." This category includes a vast array of plants grown in homes and gardens, as well as public and private greenhouses and conservatories and for commercial production. The floral industry utilizes cut flowers for making such items as arrangements and corsages. Container plants for indoor and outdoor use are also produced by this industry and include plants grown for their flowers and/or foliage.

Like other plants, flowers may be annual, biennial, or perennial. Many gardeners arrange their flowering areas (beds) by plant longevity, with annuals grouped together in one location and biennials and perennials grouped in another location. Flower enthusiasts value morphology, coloration, and fragrance when selecting species for production. The shape of the petals, their symmetry or asymmetry, and their size are morphological features of flowers. Colors range across the visible spectrum of light, although blue flowers are less common than red, yellow, or white flowers. Only a few plants with green flowers are grown because of their flower appearance. Selected cultivars of the orchid family (Orchidaceae) have green coloration on their flowers.

Some flower groups with the same common name exhibit a multitude of flower sizes, colors, and color patterns. Dahlias (*Dahlia* spp.) range from the size of pennies up to dinner plates. Their flowers may be one solid color or patterns of more than one color. Dahlia petals also vary from flat to curled edges of multiple widths. Chrysanthemums (*Chrysanthemum* and *Dendranthemum* spp.) are another example of flowers with one common name, but many morphological variations in size, color and color patterns, and petal shape.

The cut-flower industry typically relies on flowers such as roses, carnations, gerberas, chrysanthemums, tulips, lilies, sunflowers, Peruvian lilies, baby's breath, and others. Not all flowers are suitable for use in the commercial floral industry. Those plants that are most successful will have a long vase life, are fragrant, and have long stems. In addition, plants produced for the cut-flower industry must be profitable to growers; therefore, they must have many important characteristics, like producing many reliable flowers, easy to care for and grow, and easy to transport. Many of the cut flowers used in the United States are actually produced in Colombia and Ecuador and shipped overnight by plane, where they are then distributed throughout the country.

The list of plants characterized as "container plants" is much larger than those used for cut-flower production. Because container plants and flowers grown as container plants are delivered and expected to continue to be used as live plants, the choice of which plants to grow and sell can be much larger. Fewer limitations exist for a plant that is expected to live for six months or six years—as opposed to cut flowers. Container plants are often grown in greenhouses through much of the production cycle and are delivered to retailers ready to flower or already flowering. Many container plants are annuals and will be placed in the landscape for one season. Others are perennials, but in either case, consumers respond to flowering plants. One of the critical aspects of successfully growing and selling container plants is having them at their peak for the retail season. Plants that do not appeal to consumers when they come to buy may not get sold. Container plants are often grown in nurseries, even though many of them may be annual and only survive for a single season.

d. nursery and ornamental plants

Plants classified as nursery plants are grown in individual pots for one or more years before they are considered large enough to be sold. Shrubs and trees are the most common nursery plants. Most of these plants are used as focal points of landscapes on private and public properties. Nursery plants are very different from many other types of crops because their use is primarily ornamental. They may take many years to grow to maturity.

Nursery owners invest a considerable amount of time, space, and resources to ensure the success of each individual plant. In most agricultural settings, plants are managed on the basis of a field. In the nursery, every individual plant becomes much more important. In addition, because plants may be grown three, five, or ten years before they are sold, a disease or other catastrophic event that kills all of a grower's plants may cost the grower as much as a decade of work. A tomato grower will certainly not be pleased to lose an entire year's tomato crop, but every year can bring a new crop, unlike the nursery. However, an advantage to growing many types of plants in the nursery at the same time is that, if one particular species of plant experiences a failure, most of the other species will often be fine, assuming they are not also susceptible to the disease/pest/environmental condition that killed the other species.

In order to ensure that nursery plants of the same variety/cultivar are reliably similar to one another, most plants are propagated using vegetative structures. This typically involves grafting or rooting of cut branches. The resulting plants are clones of one another, since they have the same genetic composition. Nursery growers also must contend with moving plants in the nursery as they get larger (to new pots and new locations). Nursery plants will eventually be transplanted to permanent locations as whole living plants—sometimes a difficult proposition that is not a concern for a tomato grower.

The different species and varieties of plants used as nursery landscape plants are enormous. Common ones include rhododendrons, azaleas, yews, junipers, boxwoods, privets, maples, birches, crab apples, pines (and dwarf pines), ornamental grasses, hostas, and hundreds more. In addition to growing different species, there may be many different varieties, each serving a different purpose in the landscape. Thousands of different hosta varieties exist, and some collectors will pay hundreds or thousands of dollars for a new, unique, or rare variety.

e. turfgrasses

Turfgrasses are a special group of plants in the Poaceae family that tolerate being cut shorter than their normal growing height. Turfgrasses are used on home lawns, athletic fields, parks, roadsides, golf courses, commercial properties, prisons, airports, and cemeteries and cover huge amounts of the urban and suburban landscape. The desired appearance of the turf area affects selection of turfgrass species as well as management level. Roadsides are maintained using little or no irrigation or fertilization once they are established. High-use areas like athletic fields and golf courses would receive the highest level of maintenance, with optimal irrigation and fertilization provided. Different management expectations require the use of different grasses.

Approximately 50 grasses among the hundreds of grass species withstand being mown, because their apical meristems are located close to the ground below mowing heights. These meristems, called crowns, are compressed, compared to typical meristems.

Figure 19.4. Turfgrasses are a common ornamental crop, grown on golf courses, lawns and athletic fields covering millions of acres across the country.

Since mowing is stressful to turfgrass plants, no more than one third of the green tissue should be removed during a single mowing.

Individual grass plants live approximately one year until they flower. Under mown conditions, most turfgrass flowers are removed before seeds can be produced. Many people are not even aware that flowers have been present at the beginning of the growing season. Even though individual plants usually last only a year, turfgrasses are considered perennial because they are able to produce new stems before they flower.

The species of turfgrass used in a location are influenced by their tolerances to drought, pests, fertility requirements, and temperature extremes. Grasses grown in northern regions are referred to as cool-season grasses. Grasses grown in southern areas are warm-season grasses. Most cool-season grasses are established from seed, while many warm-season grasses are propagated using sod, stolons, or sprigs. Stolons are aboveground horizontal stems with little, if any, attached soil. Sprigs are chopped up pieces of sod having some soil attached.

Figure 19.5. Turfgrass sod can be installed quickly to produce a ready-made lawn. Unfortunately, turfgrasses can be delicate and can die just as quickly if improperly installed.

In the Northern United States, Kentucky bluegrass (*Poa pratensis*) is the most popular species. This grass is the major species used in home lawns and is also a sturdy grass for athletic field use. The presence of underground horizontal stems, known as rhizomes, enable rapid recovery of damaged turf when Kentucky bluegrass is included in a mixture of species or used alone. Other cool-season grass species include the fine fescues (*Festuca* spp.), tall fescue, perennial ryegrass (*Lolium perenne*), creeping bentgrass, and even semi-weedy grasses like annual bluegrass (*Poa annua*).

The most popular warm-season grass used in the southern United States is common Bermuda grass (*Cynodon dactylon*). Cultivars of this grass may have stolons, rhizomes, or both stolons and rhizomes. The aggressive spread of this grass makes it best used alone. Common bermudagrass is used on both lawns and athletic fields where sufficient sun is available. The lack of shade tolerance is the Achilles' heel of this species. The bermudagrass used for golf course greens in the South is a hybrid between common bermudagrass and African bermudagrass (*C. transvaalensis*). This hybrid (*C. dactylon* X *C. transvaalensis*) tolerates close mowing, but does not provide as smooth a putting surface as creeping bentgrass.

Other warm-season grass species include St. Augustinegrass (*Stenotaphrum secundatum*), zoysiagrass (*Zoysia* spp.), buffalograss (*Buchloe dactyloides*), centipedegrass (*Eremechloa ophiuroides*), seashore paspalum (*Paspalum vaginatum*), kikuyugrass (*Pennisetum clandestinum*), and bahiagrass (*Paspalum notatum*).

f. forests

Figure 19.6. Tree farms can be found throughout the country as a local source for annual Christmas trees.

Forests are one of the great natural resources of North America. When the first explorers arrived in North and South America, the continents seemed to be covered in forests. A scholar of early American history and a prolific illustrator, Eric Sloane, once stated that a squirrel could travel 300 miles inland without ever touching the ground as it hopped from tree to tree. Although this may or may not have been true, today's landscape is vastly different from the one that existed only a few hundred years ago.

As more Europeans arrived and settled, forests were cleared, and the lumber that was produced was used to build houses and ships. Wood was the primary heating fuel. Forests were abundant. Cleared land was used for animal grazing and crop production.

At the beginning of the 20th century, as agriculture went into decline, many of the farms that were no longer in use were reclaimed by the forests. Today, very few places on the continent contain virgin, uncut forests. One of the largest stands of uncut trees in the Eastern United States is in Cathedral State Park in West Virginia, where an ancient hemlock forest sits on 130 acres of land. Some of the trees in this small wood are as wide as 21 feet in circumference. In the Western United States, ancient redwoods are relatively common by contrast, and some trees are estimated to be over 2000 years old.

Forests were also essential to the lives of native populations, not only for wood, but also for the wildlife they supported. The forests of North America were home to many Native American populations, and these tribes utilized these forests in every aspect of their daily lives.

Figure 19.7. Many forests are unmanaged, but even these are often used for lumber, paper pulp and fire wood.

Despite the significant reduction in forested land, modern populations are also reliant on forests. While steel and plastics are commonplace in modern societies, most of our buildings are still made from lumber. Wood is the preferred material for constructing furniture, and all of our paper is produced from trees. Additionally, people still use wood for heat in many areas throughout the world, including the most modern societies.

There are many different types of forests on the planet. Any forest that is used as a human resource should be carefully managed so as not to destroy it. When lumber is harvested from forests, it is often done selectively. That is, not every tree is removed. This allows the forest to regenerate and the soil and land to remain intact. Often, new trees are planted to replace the trees that have been removed. Different areas contain distinct forest types and diverse types of trees. In North America, the Northern forest contains many types of conifers (also known as softwoods) and extends from the northernmost region of the United States into Canada. The central forest includes a very diverse number of trees and covers 30 states. Other forest types in North America include the Western coastal forest, the Southern forest, the bottom land forest, and the Western interior forests. These names describe a type of forest, not a single continuous forest.

Different trees can produce very different types of wood, which can be used for very different purposes. Often, the lumber that is produced from trees is categorized as either softwood or hardwood. As the name implies, softwoods are less firm than hardwoods. In most cases, softwoods are conifers (gymnosperms), and hardwoods are deciduous trees (angiosperms). Softwoods are commonly used for house construction and other utility purposes. Hardwoods are often used for flooring, furniture, and other related applications. The characteristics of individual species within this broad group can also vary—some hardwoods make excellent furniture, while others are better suited for pallets.

Although forests are managed much differently than other agricultural crops, they can be susceptible to insects and diseases just like other agricultural crops. The damage that may be caused can last for generations. American chestnut blight was a disease that came into the United States in the early 1900s and killed almost every American chestnut tree in North America—literally billions of trees. The hemlock woolly adelgid is an insect that attacks hemlock trees. It was introduced into the United States in the 1920s and has spread throughout much of the Eastern part of the country. In some areas, it has destroyed many trees, while other small stands of trees remain only mildly affected. Even acid rain can cause stress on plants and lead to long-term damage.

Unfortunately, controlling forest pests is nearly impossible. Because forests span such large areas, it is difficult to apply chemicals to them. In addition, the cost of making these applications is prohibitive. Finally, forests are complex ecosystems with huge amounts of species diversity. The potentially negative environmental risks of putting large amounts of pesticides into such an ecosystem are unknown. In most cases, the best way to approach pest problems in a forest system is to slowly replace susceptible species with resistant species as the susceptible plants are harvested or die. Short of these types of remediation efforts, forests will continue to evolve and change (for good or ill) as they succumb to a combination of environmental stresses, pathogens, and pests.

Credits

1. Fig. 1.1. Source: NASA / WMAP Science Team. Copyright in the Public Domain.
2. Fig. 1.2. Source: NEUROtiker / Wikimedia Commons. Copyright in the Public Domain.
3. Fig. 1.3. Source: Mhowison / Wikimedia Commons. Copyright in the Public Domain.
4. Fig. 1.4. Source: Mhowison / Wikimedia Commons. Copyright in the Public Domain.
5. Fig. 1.5. Source: USGS. Copyright in the Public Domain.
6. Fig. 2.2. Copyright in the Public Domain.
7. Fig. 2.3. Copyright in the Public Domain.
8. Fig. 2.4. Copyright in the Public Domain.
9. Fig. 3.1. Source: Larry Rana / USDA. Copyright in the Public Domain.
10. Fig. 3.2. Source: USDA. Copyright in the Public Domain.
11. Fig. 3.3. Source: George E. Marsh / NOAA. Copyright in the Public Domain.
12. Fig. 3.5. Source: Morley / USDA. Copyright in the Public Domain.
13. Fig. 4.1. Source: APPA. Copyright in the Public Domain.
14. Fig. 4.3. Copyright © 2007 by Yym1997, (CC BY-SA 3.0) at: http://commons.wikimedia.org/wiki/File:070915HK_Air_Pollution.jpg.
15. Fig. 4.4. Copyright © 2006 by Hannes Grobe, (CC-BY-SA-2.5) at: http://commons.wikimedia.org/wiki/File:CO2-variations_hg.png.
16. Fig. 4.5. Source: Jesse Allen / NASA. Copyright in the Public Domain.
17. Fig. 5.1. Copyright © 2011 by Joydeep, (CC BY-SA 3.0) at: http://commons.wikimedia.org/wiki/File:Blue-green_algae_cultured_in_specific_media.jpg.
18. Fig. 5.2. Source: Yikrazuu / Wikimedia Commons. Copyright in the Public Domain.
19. Fig. 5.3. Copyright © 2007 by Inductiveload, (CC BY-SA 3.0) at: http://commons.wikimedia.org/wiki/File:EM_Spectrum_Properties_edit.svg.
20. Fig. 5.4. Source: Tameeria / Wikimedia Commons. Copyright in the Public Domain.
21. Fig. 5.5. Copyright © 2010 by Mike Jones, (CC BY-SA 3.0) at: http://commons.wikimedia.org/wiki/File:Calvin-cycle4.svg.
22. Fig. 5.6. Source: Mariana Ruiz / Wikimedia Commons. Copyright in the Public Domain.
23. Fig. 6.1. Source: John G. Murdoch / Wikimedia Commons. Copyright in the Public Domain.
24. Fig. 6.2. Copyright © 2005 by Eric Guinther, (CC BY-SA 3.0) at: http://commons.wikimedia.org/wiki/File:Liverwort.jpg.
25. Fig. 6.4. Source: Alexander Roslin / Wikimedia Commons. Copyright in the Public Domain.
26. Fig. 7.1. Copyright © 2008 by Hendrik128, (CC BY-SA 3.0) at: http://commons.wikimedia.org/wiki/File:Tobacco-seeds.JPG.
27. Fig. 7.2. Copyright © 2011 by Stanislav Doronenko, (CC BY-SA 3.0) at: http://commons.wikimedia.org/wiki/File:Bean_germination.jpg.
28. Fig. 7.3. Copyright © 2005 by Bluemoose, (CC BY-SA 3.0) at: http://commons.wikimedia.org/wiki/File:Sunflower_seedlings.jpg.
29. Fig. 7.4. Source: LadyofHats / Wikimedia Commons. Copyright in the Public Domain.
30. Fig. 7.6. Source: Peter Coxhead / Wikimedia Commons. Copyright in the Public Domain.
31. Fig. 7.7. Source: Scott Ehardt / Wikimedia Commons. Copyright in the Public Domain.
32. Fig. 7.8. Source: Flip Schulke / EPA. Copyright in the Public Domain.

33. Fig. 8.3. Source: Franz Eugen Köhler / Wikimedia Commons. Copyright in the Public Domain.
34. Fig. 8.6. Copyright © 2011 by Zephyris, (CC BY-SA 3.0) at: http://commons.wikimedia.org/wiki/File:Leaf_Tissue_Structure.svg.
35. Fig. 9.2. Source: Dartmouth College. Copyright in the Public Domain.
36. Fig. 9.4. Source: Kevin Holcomb / FWS. Copyright in the Public Domain.
37. Fig. 9.5. Source: Franz Eugen Köhler / Wikimedia Commons. Copyright in the Public Domain.
38. Fig. 10.3. Copyright © 2008 by Sam Beebe, (CC BY 2.0) at: http://www.flickr.com/photos/28585409@N04/2851494334.
39. Fig. 10.4. Source: USDA NRCS. Copyright in the Public Domain.
40. Fig. 10.8. Source: Keith McCall / USDA NRCS. Copyright in the Public Domain.
41. Fig. 11.7. Source: Stephen Ausmus / USDA. Copyright in the Public Domain.
42. Fig. 11.8. Source: USDA NRCS. Copyright in the Public Domain.
43. Fig. 13.2. Source: Paolo Neo / Wikimedia Commons. Copyright in the Public Domain.
44. Fig. 13.3. Source: Keith Weller / USDA. Copyright in the Public Domain.
45. Fig. 13.4. Source: Keith Weller / USDA. Copyright in the Public Domain.
46. Fig. 13.5. Source: USDA NRCS. Copyright in the Public Domain.
47. Fig. 13.6. Copyright © 2005 by Formulax, (CC BY-SA 2.0) at: http://commons.wikimedia.org/wiki/File:Strawberries.JPG.
48. Fig. 13.7. Source: Jvlietstra / Wikimedia Commons. Copyright in the Public Domain.
49. Fig. 14.1. Source: Dladek / Wikimedia Commons. Copyright in the Public Domain.
50. Fig. 14.2. Source: Keith Weller / USDA. Copyright in the Public Domain.
51. Fig. 14.4. Source: Andrew McMillan / Wikimedia Commons. Copyright in the Public Domain.
52. Fig. 14.5. Copyright © 2007 by Pmg, (CC BY-SA 3.0) at: http://commons.wikimedia.org/wiki/File:Pecan-flowers.jpg.
53. Fig. 14.6. Copyright © 2010 by David R. Tribble, (CC BY-SA 3.0) at: http://commons.wikimedia.org/wiki/File:Pecans-4352.jpg.
54. Fig. 14.7. Copyright © 2010 by Istvan Takacs, (CC BY-SA 3.0) at: http://commons.wikimedia.org/wiki/File:Walnut%27s_Home.jpg.
55. Fig. 14.8. Source: USDA NRCS. Copyright in the Public Domain.
56. Fig. 15.2. Source: Gruepig / Wikimedia Commons. Copyright in the Public Domain.
57. Fig. 15.4. Source: Scott Bauer / USDA NRCS. Copyright in the Public Domain.
58. Fig. 15.5. Source: USDA. Copyright in the Public Domain.
59. Fig. 15.7. Source: Bill Tarpenning / USDA. Copyright in the Public Domain.
60. Fig. 15.2. Copyright © 2006 by Stephane8888, (CC BY-SA 2.5) at: http://commons.wikimedia.org/wiki/File:Potatoes_Vitelotte.jpg.
61. Fig. 16.5. Source: USDA ARS. Copyright in the Public Domain.
62. Fig. 16.6. Source: Martin Kozák / Wikimedia Commons. Copyright in the Public Domain.
63. Fig. 16.7. Source: Bill Tarpenning / USDA. Copyright in the Public Domain.
64. Fig. 17.2. Copyright © 2006 by Miguel303xm, (CC BY-SA 2.5) at: https://en.wikipedia.org/wiki/File:Flowers_of_the_spearmint.JPG.
65. Fig. 17.3. Source: Ranveig / Wikimedia Commons. Copyright in the Public Domain.
66. Fig. 17.4. Source: Stephen Ausmus / USDA ARS. Copyright in the Public Domain.
67. Fig. 17.5. Source: Leon Brooks / Wikimedia Commons. Copyright in the Public Domain.
68. Fig. 18.3. Source: Scott Bauer / USDA ARS. Copyright in the Public Domain.
69. Fig. 18.4. Copyright © 2007 by Aviad Bublil, (CC BY-SA 3.0) at: http://commons.wikimedia.org/wiki/File:Wheat-haHula-ISRAEL2.JPG.
70. Fig. 18.5. Source: Charlie Fong / Wikimedia Commons. Copyright in the Public Domain.
71. Fig. 19.1. Source: David Nance / USDA ARS. Copyright in the Public Domain.
72. Fig. 19.2. Source: Bob Nichols / USDA NRCS. Copyright in the Public Domain.
73. Fig. 19.3. Copyright © 2011 by Kevinbercaw, (CC BY-SA 3.0) at: http://commons.wikimedia.org/wiki/File:Tobacco_Flowers.jpg.
74. Fig. 19.7. Source: Paolo Neo / Wikimedia Commons. Copyright in the Public Domain.

CPSIA information can be obtained at www.ICGtesting.com
Printed in the USA
LVOW02s0608270813
349785LV00001B/1/P